TOURING NORTH AMERICA

SERIES EDITOR
Anthony R. de Souza, *National Geographic Society*

MANAGING EDITOR
Winfield Swanson, *National Geographic Society*

CANADA NORTH

Journey to the High Arctic

BY
J. K. STAGER
AND
HARRY SWAIN

RUTGERS UNIVERSITY PRESS • NEW BRUNSWICK, NEW JERSEY

This book is published in cooperation with the 27th International Geographical Congress, which is the sole sponsor of *Touring North America*. The book has been brought to publication with the generous assistance of a grant from the National Science Foundation/Education and Human Resources, Washington, D.C.

Copyright © 1992 by the 27th International Geographical Congress
All rights reserved
Manufactured in the United States of America

Rutgers University Press
109 Church Street
New Brunswick, New Jersey 08901

The paper used in this book meets the minimum requirements of American National Standard for Information Sciences—Permanence of Paper for Printed Library Materials, ANSI Z39.48-1984.

Library of Congress Cataloging-in-Publication Data

Stager, J. K.
  Canada north: journey to the high arctic / by J. K. Stager and Harry Swain.—1st ed.
    p. cm. —(Touring North America)
  Includes bibliographical references and index.
  ISBN 0-8135-1890-3 (cloth)—ISBN 0-8135-1891-1 (paper)
  1. Arctic Archipelago (N.W.T.)—Tours. I. Swain, Harry. II. Title. III. Series.
F1060.4.S73   1992
917.19'5043—dc20   92-11925
                                                                                     CIP

First Edition

Frontispiece: Sewing skills were highly prized in traditional Inuit society. The mother in this 1950 picture wears a caribou-skin *amautig* and pants but uses a steel needle and thimble./Richard Harrington; National Archives of Canada, PA 166824.

Series design by John Romer

Typeset by Peter Strupp/Princeton Editorial Associates

△ Contents

FOREWORD *ix*
ACKNOWLEDGMENTS *xi*

PART ONE
INTRODUCTION TO THE REGION

Canada North 3
Contact: Colliding Cultures 28
Inuit: The People 39
Canadians and the North 60

PART TWO
THE ITINERARY

Prologue		*69*
DAY ONE:	Iqaluit and Pangnirtung	*71*
DAY TWO:	Pangnirtung, Broughton Island, Isabella Bay, Clyde River, and Pond Inlet	*81*
DAY THREE:	Pond Inlet, Koluktoo Bay, Nanisivik, Arctic Bay, Prince Leopold Island, and Resolute	*93*
DAY FOUR:	Resolute, Polaris Mine, Polar Bear Pass, Bent Horn, and Magnetic North Pole	*105*
DAY FIVE:	Resolute, Beechey Island, Truelove Lowlands, and Grise Fiord	*113*
DAY SIX:	Resolute, Geodetic Hills, and Eureka	*123*
DAY SEVEN:	Eureka, Tanquary Fiord, Lake Hazen, Ellesmere Island National Park Reserve, Fort Conger, Alert, and Ward Hunt Island	*129*

CANADA NORTH

DAY EIGHT: Eureka, Resolute, Iqaluit, Great Plain of the
Koukdjuak, and Ottawa *139*
Epilogue: The Pickup Truck of the Arctic *141*

PART THREE
RESOURCES

Fact Sheet for the Northwest Territories *145*
Hints to the Traveler *147*
Suggested Readings *151*

INDEX *153*

◮ Foreword

Touring North America is a series of field guides by leading professional authorities under the auspices of the 1992 International Geographical Congress. These meetings of the International Geographical Union (IGU) have convened every four years for over a century. Field guides of the IGU have become established as significant scholarly contributions to the literature of field analysis. Their significance is that they relate field facts to conceptual frameworks.

Unlike the last Congress in the United States in 1952, which had only four field seminars, the 1992 IGC entails 13 field guides ranging from the low latitudes of the Caribbean to the polar regions of Canada, and from the prehistoric relics of pre-Columbian Mexico to the contemporary megalopolitan eastern United States. This series also continues the tradition of a transcontinental traverse from the nation's capital to the California coast.

Because of the adverse physical circumstances, most people will relish this *Canada North* guide as an insightful survey of magnificent alpine and glacial scenery and wildlife as well as of a wide range of problems besetting the high latitudes—problems including those of the indigenous populations, mining, defense systems, sovereignty claims, communication and transportation, and considerations pertaining to tourism. John Kimberly Stager, one of the field-guide authors, received his doctorate from Edinburgh University and for six years was engaged in Arctic research for the Geographical Branch of the Canadian government. In addition, he has served on the faculty and administration of the University of British Columbia for three decades. His celebrated career includes nearly 40 years of Arctic field research on topics ranging from glacial geomorphology, fur trade posts, the economics of reindeer raising, and the business of Inuit corporations. His field research has extended from the Mackenzie delta in the far northwest to the settlements of Baffin Island. Harry Sheldon Swain, the other author and also a Canadian, obtained his doctorate from the University of Minnesota and subsequently was a Canada Council Fellow at the University of Cambridge. His last two decades of distinguished public service have comprised various senior posts in the Canadian government, including his present role as deputy minister of Indian Affairs and Northern Development.

Anthony R. de Souza
BETHESDA, MARYLAND

△ Acknowledgments

We acknowledge the dedicated work of the following cartographic interns at the National Geographic Society, who were responsible for producing the maps that appear in this book: Nikolas H. Huffman, cartographic designer for the 27th International Geographical Congress; Patrick Gaul, GIS specialist at COMSIS in Sacramento, California; Scott Oglesby, for the relief art work; Lynda Barker; Michael B. Shirreffs; and Alisa Pengue Solomon. Assistance was provided by the staff at the National Geographic Society, especially the Map Library and Book Collection, the Cartographic Division, Computer Applications, and Typographic Services. Special thanks go to Susie Friedman of Computer Applications for procuring the hardware needed to complete this project on schedule.

We thank Lynda Sterling, publicity manager and assistant to Anthony R. de Souza, the series editor, and Richard Walker, editorial assistant at the 27th IGC. They were major players behind the scenes. We also thank Natalie Jacobus for proofreading the volume and Tod Sukontarak for indexing the volume. Many thanks, also, to all those at Rutgers University Press who had a hand in the making of this book, especially Kenneth Arnold, Karen Reeds, Marilyn Campbell, and Barbara Kopel.

From Ottawa, Danielle Carrière-Paris and Patricia Boucher researched photos, Barbara Henry typed much of the text, and Joanne Aubé organized both the authors and the reconnaissance trip in the summer of 1991. Bruce Howe and Bonnie Hrycyk arranged for the hospitality of the Polar Continental Shelf Project at Resolute, and Andy Thériault guided the party through Iqaluit and Apex. It would take too much space to thank each of the distinguished (and, on occasion, nearly extinguished) members of the reconnaissance party individually, but they know who they are, and their sober reflections greatly improved our recommended itinerary.

Errors of fact, omission, or interpretation are entirely our responsibility, and any opinions or interpretations are not necessarily those of the 27th International Geographical Congress, which is the sponsor of this field guide and the *Touring North America* series.

PART ONE

Introduction to the Region

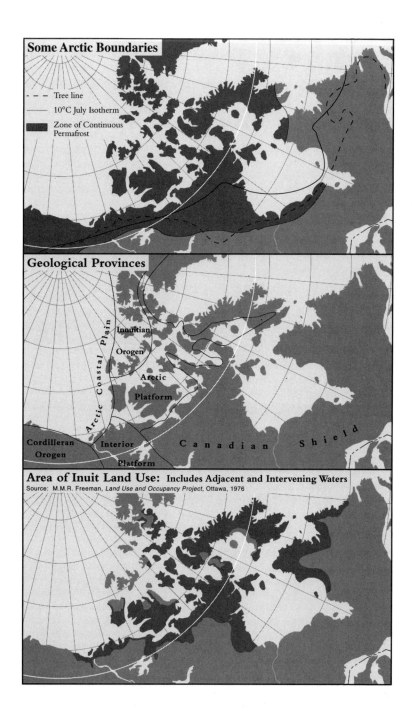

CANADA NORTH

Most people picture the far north of Canada, its Arctic, as a snow-covered land with frozen seas, a region where air temperatures place well below 0° on any scale, a terrain engulfed in darkness at least part of the year. The winter image is the popular one, carrying with it a sense of foreboding, even danger, and for the intrepid, a challenge. Outsiders regard Canada's Arctic as an exotic landscape in which the limits of nature are extreme, but also one in which marvelous adaptations have generated an intriguing array of plant and animal life. Supreme over all are the indigenous Eskimo—or Inuit, as they prefer to be known—who are, for schoolchildren the world over, the classic example of human ingenuity in response to harsh environmental dictates. As with every romantic image, the popular conception of Arctic Canada contains elements of truth amidst fantastic or outdated images that need to be adjusted in light of a pervasive modern world.

    The natural landscape endures, and is probably not very different in general appearance than when it first greeted those Europeans who opened the region to the outside world. Ice-covered seas that frustrated early explorers are today obstacles to would-be maritime travelers. Winter storms can still confine human activity as they did the Inuit in prehistory. In summer the beauty of flowers and torment of flies return regularly in their season and without fail. Human occupancy, on the other hand, is much changed. Inuit have newcomers for neighbors who share the mixed blessings of their industrial technology and entrepreneurial values. So rapid has been societal change that styles of life dependent on nature's

4 CANADA NORTH

regimes and resources and on modern infrastructure and organization are both in the experience of living persons. Casual observers of present-day Arctic settlements, with their modern housing, power, transportation, satellite communications, and structured social organization, are tempted to marvel at an indigenous people transformed. But like the land and sea surrounding it, Inuit society maintains its essential character and identity.

THE ARCTIC

Taken from astronomy, the term Arctic derives from the great bear constellation Ursa Major—called Arktos in Greek—that rotates around a fixed star above the Earth's North Pole. It is but a geometric step to identify the Arctic Circle, 66°30′N, as the boundary north of which both continuous daylight or darkness is possible because the Earth's axis is inclined to the plane of the ecliptic on which it revolves about the sun. The Arctic Circle, however, does not delimit the Arctic environmental zones—zones with physical and biological characteristics that are Arctic in nature.

The dominant visual image of northern Canada is its vegetative cover. You enter the north when you enter the boreal forest. This broad zone, up to 800 km (500 mi) wide, sweeps from Yukon and the Alaska panhandle across the country to the northern edge of Lake Superior and the southern shore of James Bay, to reach the Atlantic coast north of the St. Lawrence estuary on the Labrador Sea. Except where it is interrupted by bare rock or lakes, a close-crowned forest of needle-leaf, coniferous trees forms this habitat of migrating woodland birds, who feed in the short summer on the biting insects that are the torment of other animals. In winter, the forest is a trap for falling snow that piles deep and soft, protected from transporting winds.

Parallel to and north of the boreal forest is a zone that, because of low temperatures and a short growing season, is open woodland. It hosts the same species as the boreal forest, but the trees grow in clumps, copses, and protected valleys and decrease in density and number in a transition zone toward the northern edge. The open woodland is a band 300 to 500 km (200 to 300 mi) wide that reaches almost to the Beaufort Sea at the mouth of the Mackenzie River, touches south shores of Hudson and James bays, then crosses the Ungava Peninsula to the Atlantic coast.

The northern border of the open woodland is the tree line, beyond which trees do not grow. Poleward of the tree line is the tundra zone, a quarter of Canada, a region in which plant life, where it exists, forms a carpet of vegetation close to the ground. There is no shelter from the wind, and in winter the snow, driven and drifting across the surface, is packed hard into ridges and ripples that betray the wind's prevailing directions. Tundra land is Arctic land and indeed, for many purposes this is an adequate definition. In summer when the snow disappears, the variety of life that subsists on the tundra emerges. The plants and animal species here are restricted, however; the tundra is less biologically productive than the zones to the south. On the mainland and southern parts of Canada's northern islands are well-vegetated, grass- and heath-covered surfaces. Farther north, with a shorter summer and drier climate, vegetation—mainly moss and lichens—becomes patchy and bare rock is exposed. The far north of the Queen Elizabeth Islands is mostly tundra desert. Dry, rocky surfaces, sometimes with spectacular lichen blooms, support sparse plant cover near intermittent streams or snowbank seeps. Autumn on the vegetated tundra, late in August and early in September, briefly produces a carpet of rusts and browns before the white of winter returns.

THE LAND

Arctic land surfaces owe their character, in part, to the underlying geology. Four known geological provinces underlie the Canadian Arctic. The oldest rocks with the longest and most tumultuous

Auyuittuq Park Reserve. The granite cliffs of this recently deglaciated area tower a vertical mile above the valleys. Photograph by H. Swain.

geologic history are found in the Canadian Shield, the foundation of the North American continent. Composed of granites, gneisses, ancient sediments, and volcanics, the Shield in the north is exposed on the Ungava Peninsula and the southern end and northeastern two-thirds of Baffin Island. The Shield underlies the surface on the eastern end of Devon Island and extends north along the eastern edge of Ellesmere Island halfway to the northern tip. In this location, the Shield becomes the high eastern rim of the Canadian Arctic. The elevations of its mountainous forms increase northward from 1,000 m (3,000 ft) on South Baffin Island to over 1,800 m (5,900 ft) on Ellesmere Island. The surfaces supported by this eastern half of the Shield bear all the traces of Pleistocene glaciation, particularly well-developed alpine forms, including one of the most beautifully fiorded coasts in the world. Except where remnant icecaps cover the high ground, much of the land is exposed bedrock. West of Hudson Bay lies a lower, level-to-undulating section of the Shield, the boundary of which intersects with the Arctic coast at Coppermine. There are two fingerlike northward extensions of this Shield: one forms the backbone of Boothia Peninsula and western Somerset Island, and the other incorporates Melville Peninsula and crosses to Baffin Island. This Shield country is bare rock, either worn down and glacially scoured or lightly mantled by glacial drift. Remarkable are the thousands of lakes, poorly integrated into an immature drainage pattern, that fill the depressions caused by structures and joints of ancient formations, or by recent erosion and deposition. Viewed from on high, the Shield could fairly be described as having a 'beefsteak' topography. Surface elevations increase from Hudson Bay westward to the outer Shield margin where its edge—a kind of raised rim—is 300 to 600 m (1,000 to 2,000 ft) above sea level. Broad, domelike rises, over 300 m (1000 ft) high, occur on central Boothia Peninsula, on Melville Peninsula, and where Boothia and Melville join the main body of the Shield, northwest of Hudson Bay. Elsewhere the surface elevations remain between 100 and 200 m (300 and 700 ft) above sea level.

    Overlapping and up against the Shield is the Arctic Platform, a broad, geologically stable zone of flat-lying sedimentary rock. The

Platform is built layer upon layer, and, where it is exposed, cliffs and hillsides show successive layers of limestones, dolomites, shales, and other sediments. The Arctic Platform underlies the islands between the mainland coast and the Parry Channel, extending north of the channel over Cornwallis and Devon islands. In the east, the Platform surrounds the Brodeur Peninsula, the 'crab claw' on the northwest corner of Baffin Island, then squeezes along Baffin's southwest coast and cuts across Foxe Basin. The Arctic Platform land surface features some rock outcrops but is mainly composed of weathered bedrock and glacial deposits. Elevations are low; most of the Platform stands under 100 m (300 ft), but high areas in the west and northeast rise to 500 m (1,600 ft) and higher. Coastlines of cliffs can be seen along Lancaster Sound, and, where there is high ground, local relief of 200 m (700 ft) or more is common.

The Innuitian Orogen geological province covers the best part of the Queen Elizabeth Islands. It is a mountain-building province where accumulated sediments have been folded, faulted, and elevated at the edges of the Shield and the Platform. From the northern half of Ellesmere Island, in a broad belt that lies south and west across the islands, sub-parallel folding gives a grain to the landscape, which is oriented in successive ridge or mountain lines that run roughly northeast to southwest (i.e., parallel to the axis of the belt). The western two-thirds of the province is covered mainly with unconsolidated materials, partly glacial, and elevations run in the 100- to 200-m (300- to 700-ft) range, although Victoria Island and Banks Island sport some hills that top 300 m (1,000 ft). The eastern third has more bedrock exposed, and, because little or no vegetation is present, the folded structures form almost textbook examples of synclines and anticlines, in rock that is often quite colorful. In this mountainous part of the Orogen, peaks above 1,100 m (3,600 ft) are often covered by icecaps. The glacier summits rise to heights of 2,000 to 2,900 m (6,600 to 9,500 ft). Alpine glacial landforms with long, straight trenches, hanging valleys, and active glaciers reaching into coastal fiords offer unsurpassed scenery.

The Arctic Coastal Plain is the narrow seaward edge of the low islands in the western High Arctic. Its level sediments slope gently into the Beaufort Sea, where it is still being extended as it receives the products of erosion and as isostatic rebound continues to lower sea levels. Nowhere do these unconsolidated, mainly alluvial land surfaces rise far above sea level.

WEATHER AND CLIMATE

Most of the time and over all of its area, the Arctic is cold. It is a global heat sink, a low-energy environment that produces mean annual temperatures around or below freezing and winters that are spectacularly frigid and long. The basic cause of these conditions is the pattern of energy exchange. In summer, when most of the energy is incoming, continuous daylight begins after the spring equinox at the North Pole and spreads south to reach the Arctic Circle—66°30′N—on the summer solstice; then the progression reverses until the sun disappears below the horizon at 90° N with passage of the fall equinox. The Arctic receives most of its annual energy allotment—solar radiation—during these six months when daylight is available. However, as latitude increases, solar radiation becomes more and more attenuated, because the sun does not rise high in the sky and the long, oblique trajectory of the sun's light through the atmosphere reduces the amount of energy that reaches the surface. In winter the situation is reversed so that most high-latitude radiation is outgoing. Darkness increases and energy loss intensifies from late September until a peak in late December, then darkness and energy loss simply continue until near the end of March. Thus, despite abundant daylight opportunity, the atmosphere is only weakly heated in summer, and in the winter it is profoundly cooled. Were it not for general atmospheric circulation and ocean currents, the polar regions would grow colder and colder.

Air circulation patterns over the region determine the Arctic climate. Although daily and weekly variations occur, a core pattern of wintertime air flow relates to a broad zone of low pressure with its counterclockwise circulation centered south and east of Greenland. The companion high-pressure cell of dense, cold air forms a shallow mass over the Mackenzie lowlands and nearby Arctic Ocean. Both pressure systems contribute to a persistent outflow of cold Arctic air from the snow and ice-covered polar basin, southeast across the archipelago, mainland, and Hudson Bay, and into eastern Canada.

This pattern effectively prevents any incursion of warmer air and confirms the popular image of the frozen Arctic. Mean daily temperatures in midwinter range from –22°C to –27°C (–8°F to –17°F) in northern Ungava and along the Baffin Island coast. In the lower Mackenzie valley and over the Arctic archipelago the temperatures average –30°C (–22°F). The Canadian Arctic is coldest in the District of Keewatin, where mean daily temperatures range from –32°C to –35°C (–26°F to –31°F). These are average temperatures; extremes of –40°C (–40°F) and lower can persist for days and even weeks.

Intensifying the cold is the persistent wind of the central Arctic. In the core region comprising Keewatin, northern Hudson Bay, and Foxe Basin, the wind blows over 90 percent of the time, with average speeds of 20 km/hr (12 mph), and usually from the NNW direction. Wind blows in eastern Baffin Island 80 percent of the time, with velocities averaging around 15 km/h (9 mph). The outer margins of the high Arctic islands experience calm up to 30 percent of the time, and winds there are weaker, with a 10 km/h (6 mph) mean. The wind speeds are stronger in winter and, combined with low temperatures, produce marked wind chill. In terms of temperature equivalents, wind speed of 20 km/h (12 mph) makes –10°C (14°F) air feel like –14°C (8°F) and –30°C (–22°F) feel like –39°C (–38°F). If that wind speed were to double, –10°C (14°F) and –30°C (–22°F) would feel like –21°C (–6°F) and –51°C (–60°F), respectively.

Actual examples can be dramatic. Iqaluit in February of 1979, for instance, had twelve days of –41°C (–42°F) air accompanied by 60 km/h (37 mph) winds with gusts of 100 km/h (62 mph). In 1975, Hall Beach experienced –46°C (–51°F) with winds of 61 km/h (38 mph), for a wind-chilled temperature of –86°C (–123°F).

Summer circulation sees the low-pressure cell near Greenland weaken and center itself over Baffin Bay and Hudson Strait. Into the cell pours a regular procession of cyclonic storms that have crossed the continent on tracks over the northern boreal forests. They bring stormy weather to the eastern Arctic. The winter high-pressure cell of the western Arctic retreats altogether and lies weakly over the polar basin. With the rapid return of lengthening days, the snow cover melts and both lake ice and sea ice disappear, except in parts of the High Arctic archipelago and on the Arctic Ocean. The melting process consumes a significant portion of the energy available in spring. Arctic air never warms very much, even with continuous daylight, because of the large areas of frigid open ocean. The Mackenzie basin is an exception, because the distance from cold seawater permits the build-up of higher temperatures fairly far north. Almost all of the Arctic has a mean temperature of less than 10°C (50°F) in the warmest month, July. In fact, the 10°C July isotherm is often accepted as the boundary between the sub-Arctic and the Arctic. Interior locations such as Baker Lake average above 10°C (50°F), but along the mainland coast and into Hudson Bay mean summer temperatures at settlements range from 6°C to 8°C (43°F to 46°F); the warmest part of the day may be 3°C or 4°C higher. The Baffin coast and southern archipelago average 5°C (41°F), and the extreme north has a midsummer mean of 4°C (39°F). Record hot days have reached 30°C (86°F) at sites away from the coast, but temperatures on the coast rarely exceed 20°C (68°F). (Rankin Inlet reached 30.5°C [86.9°F] on 9 August 1991, an all-time record. Inuit children swam in a tarn atop an esker, while well-prepared southern visitors sweated in their long johns.)

The Arctic is one of the driest regions on the globe. Much of the area lies outside the usual pattern of cyclonic activity, and, in any case, cold Arctic air does not hold much water vapor. Annual precipitation ranges between 200 and 500 mm (8 and 20 in.) over the mainland and south Baffin Island. In fact, precipitation is high on south Baffin, but decreases north along the mountain fringe as a result of less frequent cyclonic activity in this low-pressure zone. Most of the Arctic islands, however, expecially those in the north and west, are very dry, many receiving less than 100 mm (4 in.) of precipitation annually. Sparse vegetation and large expanses of bare rock and soil attest to true desert conditions. Rainfall is concentrated into the four months of summer and accounts for 40 to 50 percent of the annual precipitation, except in the shorter summer of the High Arctic, when rainfall makes up about 30 percent of the total precipitation. Despite the small concentrations rain falls about one summer day in four, and more frequently in Hudson Strait. Snow, however, dominates the northern climate. It covers the ground about ten months of the year, from September 1 until nearly July 1 on the northern islands, from September 15 until June 1 on the main islands and Baffin, and from late October until early May on the mainland in the northwest. Snowfall is light on the islands, 60 to 100 cm (24 to 40 in.), but can reach 200 cm (80 in.) elsewhere. The heaviest snowfall occurs along the east coast of Hudson Bay and on south Baffin Island. The relatively late freezing of Hudson Bay provides moisture for the westerly march of cyclonic storms over Ungava and south Baffin.

SPECIAL WEATHER EFFECTS

Winter

Winter, as we know, is dominated by cold, snow, and darkness. It is also accompanied by certain weather circumstances that pose risks to people and animals. Wind and snow combined determine the nature of the ground surface for all the tundra and sea ice. Although snow falls in light amounts, it is very fine and blows easily. The surface can become hard-packed with ripples and longitudinal ridges betraying the prevailing wind direction, or snow may drift into soft dunes in the valleys or lee of some obstruction. Accumulations of snow are constantly reworked by the wind and the snow blanket is usually thin, except where drifts form. The Inuit have many words for snow that describe its shapes and densities; each word has meaning for people who use snow for navigation and travel, for understanding the behavior of game, and to build shelter. Blowing snow is the chief cause of poor visibility in winter, and it is common. Resolute and Baker Lake report an average of ninety days a year with these conditions, and the persistence of such storms effectively confines people indoors.

Spring and Fall

In spring and fall, when the sun is low in the sky, cloudy conditions can produce a uniform whiteness to land and sky that is unbroken by shadows. The horizon seems to disappear. This condition, known as a whiteout, makes travel dangerous, because without shadows no sense of topography or scale is possible—one can easily fall by stepping or driving off a rise or small cliff.

At the edge of the sea, or on the sea ice if open water is exposed in cracks or leads, the juxtaposition of warm sea water and very cold air produces an ice-crystal fog: the Arctic sea smoke. This local phenomenon persists in calm conditions over open water. The Inuit locate open water by sea smoke and can exploit it for hunting or avoid it when traveling.

Another weather feature is the result of the frequent occurrence of temperature inversions that contain very cold air next to the ground. In settlements and larger towns where heating vents, chimneys, and vehicle exhaust are prevalent, quantities of water vapor are emptied into the air and freeze to form an ice-crystal fog that reduces visibility and may make breathing difficult. Rising temperatures and wind are needed to clear the fog.

Summer

Summer winds generate waves that can restrict water travel. Visibility along the coasts, where most people live, is reduced by fog and cloud cover, which accompany open water especially when relatively cold water comes into contact with warmer air over the land. Light winds allow fog to form, and then, if the winds strengthen, fog banks can be blown inland and become low stratus clouds. On average, as many as ten days a month in summer are foggy. Were strong winds less common, fog would be even more persistent.

CLIMATE CHANGE

There is general agreement that the global atmosphere has warmed by about 0.5°C in the twentieth century. Climate simulators, refining and running their models based on increases in carbon dioxide and other greenhouse gases, predict that the Arctic regions will be most affected by rising temperatures. Evidence shows that, during the first half of the twentieth century, warming was noticeably stronger in the Arctic than farther south. Temperatures in the Arctic increased again during the last three decades, but the effects of the warming are not uniformly experienced in all parts of the Arctic. For example, northwestern Canada has warmed, while southern Greenland is colder. In a recent study, however, it is reported that a decade of satellite monitoring has recorded a shrinking of the area of Arctic sea ice and a decrease in the open water within it. In another paper it is revealed that sea ice is thinning. Some scientists have also predicted the retreat of permafrost. Models of climate change, despite modifications to account for new thinking and new data, continue to predict global warming, and as yet no evidence discredits the notion that the Arctic will be the first region to show the effects. Just what those effects will be is the subject of a good deal of speculation. Predictions based on characteristics of the Arctic environment include permafrost decay, with its engineering consequences; changes in sea level; increased clouds and precipitation; and altered ranges for vegetation and animal life, both land and sea. Considerable debate continues about what is really happening in the Arctic and what the future will prove. The lack of sufficient long-term data exacerbates disagreement. Nevertheless, the Arctic could well be the place to watch for the signs of global warming's true consequences and for possible solutions.

PERMAFROST

A distinctive physical condition, by whose margins the Arctic is further defined, is continuously frozen ground or permafrost. By minimum scientific definition, permafrost is ground that remains under the temperature of 0°C (32°F) continuously for more than two years. Usually, however, permafrost persists for hundreds of years, and in some areas of Canada it may be one hundred thousand years old or older. Because permafrost is defined by temperature, it is best described as a climatic condition. Changes in climate will, therefore, cause changes in permafrost.

Permafrost underlies half the area of Canada. Two geographical zones of permafrost are recognized, namely discontinuous permafrost, which underlies mainly the boreal forests, and continuous permafrost, in the tundra zone. In continuous permafrost regions, permafrost is found everywhere under the ground except where bodies of water are so deep that they do not freeze to the bottom in winter and so broad that freezing cannot reach under from the sides. The southern boundary of continuous permafrost closely approximates the location of the July 10°C isotherm and especially the tree line west of Hudson Bay, but east of the bay, the boundary lies farther north, crossing the tip of the Ungava Peninsula and south Baffin Island through Iqaluit. Thus, most of the Canadian Arctic has continuous permafrost.

Permafrost depths are determined by the intensity of sub-zero (Celsius) air temperatures and the length of time that freezing persists. In the western Arctic and on some of the High Arctic islands where conditions were too dry to generate icecaps, the land surface has been exposed to below-zero (Celsius) air temperatures for tens of thousands of years, and at sites such as Prudhoe Bay, Alaska, permafrost can be 600 m (2,000 ft) thick. At Resolute it reaches 400 m (1,300 ft) down, at Bent Horn on Cameron Island 700 m (2,300 ft) and at Alert 1,000 m (3,300 ft). In unusual circumstances, permafrost exists under the sea. It may be residual from previously dry, frozen land that submerged, or it may be generated by sea temperatures persistently just below 0°C (32°F).

16   CANADA NORTH

Bent Horn tundra truck. Transporting oilfield equipment is easy in winter, but in summer special equipment is required to avoid damaging—or bogging down in—the permafrost active layer. Despite heavy loads, this buggy's footprint weighs only four pounds per square inch. Photograph by H. Swain.

Above the land surface, the seasonal range of air temperatures at all places in the Arctic moves above freezing for some period in the summer, and this warming engenders melting of the top layer of ground. This layer of ground, known as the active layer, melts and refreezes most years, and the zone thickens from north to south. At any one location, however, the depth of the active layer can also vary greatly within a short distance, and the amount of summer melt depends upon a range of physical conditions at the micro and macro scale that affect the thermal balance from the air as well as the ground.

Where water was in the soil when permafrost was formed, the water is present as ice. Ground ice ranges in form from small crystals that bind soil particles together, to lenses and layers up to tens of meters thick. Some types of ice formed after permafrost developed and are related to certain landforms. Wedge ice, for example, occupies nooks below the cracks of tundra polygons, and ice cores are at the center of blisterlike hills called pingos. Ice in permafrost is stable as long as it remains frozen, but any circumstance that alters the thermal balance can cause surface collapse, rapid erosion, and damage to structures built on the permafrost. The role of ice in permafrost is the subject of considerable scientific scrutiny, since understanding that role is crucial to solving many engineering problems.

The surface of the ground in permafrost regions frequently displays patterns of hummocks, circles, polygons, nets, steps, and stripes. In many cases, the clastic material is sorted by size, and thus sorted circles, polygons, and the like are to be seen. All are affected by freeze-thaw action which, in the presence of water and ice, introduces physical force influenced by gravity and moving water. For the most part these processes operate in the active layer. Except for exposed bedrock, practically all tundra surfaces have some patterned ground.

CURRENTS AND SEA ICE

The general circulation of the Arctic basin close to its shore is counterclockwise. Warm waters from southern latitudes enter the basin mostly from the North Atlantic, south of Spitzbergen, Norway, and to a lesser extent from the Pacific through Bering Strait. In addition, summer surface drainage pours fresh water into the Arctic Ocean from several major Russian rivers as well as the Mackenzie River system in Canada. The rivers, incidentally, flush quantities of driftwood into the basin, where the wood is distributed by shore currents well beyond its sources. In the past, driftwood was a highly prized material among the Arctic's aboriginal people.

Part of the North Atlantic Drift curves west and south to wash the east coast of Greenland and then flows north along the west Greenland shore where it is warm enough to maintain a prolonged open water season beyond 75° N, and an all-winter one to 65° N. In north Baffin Bay the drift recurves west and sets south as the cold Labrador Current, carrying the calved icebergs from the ice face of the Greenland Icecap and the fiords of eastern Ellesmere, Devon, and Baffin islands. Part of this water, with its icebergs, washes the Labrador coast, presenting the same danger that led the *Titanic* to sink off Newfoundland on 15 April 1912, with a loss of over 1,500 lives. (The *Titanic*, then the largest ship afloat, was on its maiden voyage, Southampton to New York City, and carried over 2,200 people.) The rest of this current swings west into Hudson Strait, hugging the north shore, and enters Hudson Bay, joining a counterclockwise gyre that draws it throughout the bay. Returning water passes out of the strait along the south shore.

North of Alaska in the Beaufort Sea, a large gyre circulates clockwise, with currents passing south along the western edge of the Canadian archipelago. Some of the water entering the Arctic basin through Bering Strait heads eastward along the north slope of Alaska and across the Mackenzie mouth into Amundsen Gulf. The counterflow produced by the eastward-flowing shore drift and the westward-moving gyre produces dangerous ice conditions in this shallow offshore area. Within the Arctic islands themselves, water generally flows from the Arctic basin through the island channels southeast to feed the Labrador Current.

The Arctic seas are covered with ice. Shifting, grinding, and growling, pack ice circulates all year on the Arctic Ocean and, in winter, on Hudson and Baffin bays and in Lancaster Sound. Most of the floes in the pack are more than a year old and are two to four meters (7 to 13 ft) thick. Multiyear ice has been measured up to eight meters (26 ft) thick. Near the southern edge of ice cover, the ice is mostly first-year ice with thicknesses up to two meters (7 ft). In response to winds and current, the pack can thin or separate to expose leads of open water or be driven together into rafted piles of riprap ice 20 to 30 m (60 to 100 ft) or more thick. Pressure ridges become formidable barriers to surface travel and hang like ghostly curtains into the ocean below.

The polar pack rarely retreats from the northern edge of the Arctic islands, and old and multiyear ice chokes most of the island inlets and channels between the islands of the Queen Elizabeth group throughout the year. There are exceptions, such as Eureka Fiord, which usually opens by August. In favorable summers, open water is exposed from the Parry Channel northward and some side channels may be navigated. In Lancaster Sound and Baffin Bay, the currents, higher tides, and wind clear practically all of the ice every year, except for the stream of icebergs. These bodies are thus closed in winter mainly by first-year ice. Next to the mainland, ice clears from west to east, although sea ice frequently persists even in a good summer near Pelly Bay and west of the Boothia Peninsula. In a cold summer, ice can block the channels west to Victoria Island. The main part of the Beaufort Sea is covered with multi-year ice throughout the year.

There are areas where open water occurs at the same place every winter. These areas, called polynyas, form in response to the combined actions of wind, current and tide, and upwelling in the water column. Linear zones of open water can be found where the circulating polar pack abuts the more stable ice that locks the islands together, for example, along the shore margins of the Beaufort gyre and, similarly, along the edges of the Hudson Bay and Baffin Bay gyres. Other polynyas are lakelike expanses of open water, where currents or wind sweep ice from the sea surface. We find them at the mouths of Frobisher Bay, in Cumberland Sound, between Southhampton Island and the west shore of Hudson Bay, and at the entrance to Lancaster Sound. Small, local polynyas exist near Prince Leopold Island, in Jones Sound, and in Fury and Hecla straits. The best known polynya is the North Water, a large, persistent open sea between southern Ellesmere Island and Greenland that was discovered by whalers and used as their route to the west side. Polynyas attract sea mammals such as seals, walruses, and whales; some sea birds; and predatory polar bears. Humans, at the top of the trophic chain, have exploited the good hunting of these abrupt boundary regions for thousands of years.

The polar bear's subcutaneous fat, and white fur that conducts solar radiation down to a black, absorbent skin, help provide for survival in cold water and colder air. Photograph by DIAND.

Gregarious walruses sort themselves by sex and congregate at favorite haul-outs year after year. Photograph by DIAND.

Sea ice first forms as a floating mass of ice crystals so thick that it has a greasy appearance. Soon it sets and forms a wet, black (transparent), rubbery surface that flexes with sea swells; and even when it is thicker and gray-colored, it can bend under the weight of a passing person or sled. Wind breaks the ice apart and crushes it together, so that leads and ridges form. One-year ice formed during a long, cold winter can be two meters (7 ft) thick. If the ice does not melt completely in summer, the brine washes out. Refreezing makes multi-year ice thicker, harder, bluer, and eventually fresh enough to use for drinking water. Major accumulations of essentially freshwater ice form ice shelves, such as those growing along the north coast of Ellesmere Island. Ward Hunt Island, a frequent land base for over-ice travel to the North Pole, is surrounded by a shelf that has been measured up to 15 km (9 mi) wide. Stationary sea ice is thickened by melting and refreezing, and by accumulating snow. There is also a slow process of pure ice freezing to the bottom of the shelf where pure ice crystals are added in a sub-zero temperature environment. The shelf has a thickness of up to 45 m (150 ft) and exhibits an undulating surface with ridges that run parallel to the dominant west wind. Chunks of such shelves have broken away, becoming the origin of the ice islands of the polar basin, which because of their mass—one reached 125 sq km (48 sq mi) in area—can remain in circulation for years. Scientific investigation teams from several nations have used ice islands for base camps.

Knowledge of the properties and behavior of sea ice is extremely important to the Inuit. Certainly before contact with white explorers and whalers but even today, Inuit who live, travel, and hunt on the sea ice must understand perfectly its advantages and risks. Furthermore, the Inuit must know intimately how sea mammals adapt to sea ice if they are to exploit this source of wealth. Along the shores and in protected bays, ice forms a continuous and undisturbed surface known as fast ice, which makes for good traveling. At shorelines with high tides, the ice-foot may be frozen solidly to the land, but the tidecrack zones produce jumbled blocks of ice, replete with slush-filled holes hidden by snow. These areas are dangerous and difficult to cross. The floe edge is the seaward margin of fast ice and, because it occurs next to open water, it is an important hunting area during all the seasons when ice is present.

PREHISTORY

The only human beings to inhabit the Canadian Arctic were Inuit and their forebears, at least until contact with whites. Archaeological evidence permits the conclusion that Paleo-Eskimo hunters occupied sites on the mainland coast and adjacent islands and reached the High Arctic and northwest Greenland by 2000 B.C. It is believed that these settlers had originated to the west—perhaps in Siberia—and moved fairly rapidly along sea routes opened by retreating ice sheets in a climate that was warmer than today's.

The site of the earliest clues yet discovered of houses and lithic tools found on the upper raised beaches has been named Independence I. An Arctic small tool tradition has been identified from these sites, composed of such artifacts as microblades, scrapers, knife and weapon points, side blades, and chipped flakes of stone. At lower elevations on the beaches, and therefore probably representing later occupancies, are camp sites designated pre-Dorset. Carbon dating of remains suggests that pre-Dorset settlements appeared at least 300 years after Independence I camps, and distinctive bone and ivory harpoon heads, among other tools, mark a different culture. Pre-Dorset evidence is widespread, found mostly at coastal locations but also at interior sites south of Coppermine. Sea mammals such as seal and walrus are represented in bone and ivory. Other artifacts reveal that musk ox, caribou, polar bears, and birds were taken from the land. Harpoon heads, lances, and bows with arrows served as the basic hunting tools, and were adapted for particular game. Pre-Dorset people chipped and flaked stone, carved and polished bone and stone, made knives and scrapers, and burned fat- or oil-soaked bones and wood on flat stone stoves. Oval stone rings indicate where summer skin-tent dwellings or round-shaped stone winter houses with central passages stood.

Part of the annual round of traditional activities for coastal Inuit was the trapping of anadromous char in stone weirs. This excellent fish was dried or frozen for winter consumption by people and their dog teams. Photograph by DIAND.

Some sites include signs that the people were accompanied by dogs.

Dorset culture, which emerged from pre-Dorset between 800 B.C. and 300 B.C., seems to represent a technology evolution. An increasingly cold climate necessitated hunting through the ice. The tools used to hunt seals grew more sophisticated: double-holed, closed-socket harpoon heads for both seals and walrus; whalebone or ivory sled shoes; antler ice creepers; and snow knives. People traveled in small groups and hunted and lived on the sea ice, where the snow house, or *iglu,* may have originated. Bows, arrows, and dogs are no longer present, and it is likely that caribou, fewer then because of the colder climate, were hunted in summer at water

This Inuit family, in a picture taken in the 1950s, enjoys fresh country food beside their summer house, a skin tent. Note the skillful use of the ulu *by the lady on the left. Skin clothes and houses are rare now, but country food is still a vital part of today's diet, and a better tool than the razor-sharp, multi-use* ulu *has yet to be invented. Photograph by DIAND.*

crossings where they could be killed by lances. Dorset people built fish weirs and speared fish. Their use of fire remains mysterious; few fire-making tools have been found, and the stone lamps or pots discovered are quite small. Tent rings exist, as does evidence of sod pit-houses used in winter. The remains of stone structures reveal interiors divided into sleeping areas and a central platform reserved for alimentary functions.

An interesting feature of Dorset culture is its art. Trinket-sized renditions of sea and land animals and birds were carved of bone, ivory, wood, and stone; some pieces were scored to simulate hidden skeletons. These items apparently had both playful and supernatural functions. Representations of humans also exist and include dolls, small masks fashioned with remarkable detail, and even life-size wooden masks. Some figurines have been found marked and colored in ways that suggest their use in magic.

Dorset culture lasted until about A.D. 1400. It may have spread from a core zone comprising Victoria Island, Keewatin, Baffin Island, and Ungava to the periphery of the Queen Elizabeth Islands, Greenland, and Labrador late in its tenure. Dorset culture disappeared shortly after newcomers arrived; what exactly happened to the Dorset people is still debated. In any case, the arrival of a new people ended nearly 3,000 years of cultural continuity.

The newcomers were the Thule people. The Thule originated in Alaska and they have the hallmarks of a new breed of sea mammal hunters. When a warming climate began to reduce the pack ice in summer and prolong the season, the Thule rapidly migrated east, following the general route of the Parry Channel Northwest Passage all the way to Greenland soon after A.D. 1000, where they met newly settled Vikings from Iceland (the first Europeans to "discover" America). For a time the Thule lived alongside the Dorset inhabitants, remembered in Inuit lore today as the Tunnit, small people who were reputed to be very strong. A second pulse of migration, A.D. 1200–1300, moved south into the islands, onto the mainland coast, around Hudson Bay, and into Ungava and Labrador.

Several features of Thule life were brought intact from Alaska. Skin boats, both single-person kayaks and the larger, open *umiak,* were used for hunting. The *umiak* was also capable of moving whole groups of people and their gear from one site to the next. The Thule were skilled tool-makers who used bow drills and lacing and created a range of specific-function instruments that attest to their adaption to a mobile hunting life. The Thule people developed a whaling culture using kayaks and umiaks for transportation and harpoons with toggle heads tied to skin bladders as weaponry to capture the bowhead whale (*Balaena mysticetus*). This was a warmer period than the current one, and bowheads ranged from Baffin Bay to the Beaufort Sea and throughout the interisland channels. The Thule hunted polar bears with the help of

The wonderfully pleistocene-looking musk ox (biologically a goat, not a bovine) sheds its winter coat each summer. Skeins of musk ox wool called Qalijaqtuq *have been gathered by Inuit and other wise travelers for insulation, padding, pillows, and the like for many years. Photograph by DIAND.*

dogs and harnessed dogs to pull sleds. They learned from the Dorset the use of the *iglu,* which added flexibility to winter travel. Thule semi-permanent winter houses, constructed of stones, whale bones, and sod, featured a unique cold air trap at the entrance to prevent heat loss. They invented snow goggles and brought back the bow and arrow for hunting caribou and musk ox. They used throwing boards, bird darts and bollos, three-pronged fish spears, fish jigs and weirs, large stone lamps, and some pottery. Sewn clothing of skins resembled closely the garments seen by European explorers.

After A.D. 1200, the climate began to cool. Over the next 500 years, increasing summer ice likely had an effect on the ecology of

Arctic marine mammals, preventing bowhead whales from frequenting the archipelago channels and leaving this environment to seals. The Thule retreated with the whales, abandoning the Queen Elizabeth Islands. They turned away from whaling to concentrate on hunting walrus, seals, belugas, and narwhals, with more attention given to caribou and other land animals that were in greater numbers farther south. By the time of European contact, the changing seasonal pattern of marine and terrestrial game hunting, the accompanying tools and technology, and the summer tents and winter *iglus* or stone and bone *qarmats* signified the end of the Thule period and the transition to the historic Eskimo culture first described by explorers and others as they opened this unique realm to the outside world.

CONTACT: COLLIDING CULTURES

EXPLORATION

Commercial interest in breaking the Spanish and Portuguese dominance of Far East trade led English merchants to seek a northwest passage to the Orient in the sixteenth century. Martin Frobisher made contact with Inuit in 1576 at Frobisher Bay, and ten years later John Davis found Davis Strait and Cumberland Sound. In 1616 William Baffin pushed much farther north to Smith Sound and concluded that no passage would be found by pursuing that direction. Davis earlier had noted the possibility that Hudson Strait could be the passage and this entrance was charted by Henry Hudson in 1610. Hudson followed the bay to its southern end, where he was abandoned in 1611 by a mutinous crew who only saved their own lives by bringing home Hudson's charts and new hope of a passage west. Twenty years later, with the exploration of the south, west, and north shores of the bay by Jens Munk, Thomas James, and Luke Foxe, major doubts set in that this way led to the "Straits of Anian" and a new trade route to Asia.

Exploration was dormant thereafter until success in the fur trade (principally beaver) by the French in North America attracted English envy. Two wily and experienced back-country voyageurs, Pierre Radisson and Médard Chouart des Groseilliers, persuaded English backers to mount an expedition to the fur country through

Hudson Bay. In 1668–1669 one of two ships that started out, the *Nonsuch,* wintered over in James Bay, completing a highly successful trading mission, which led to the incorporation of the Hudson's Bay Company (H.B.Co.) in 1670, an enterprise that is still in business.

Northern exploration from this point on was motivated by the fur trade. The monopoly of the H.B.Co. was consolidated after Canada became British, and the company's efforts to explore were grudgingly pursued, more because of its charter obligations than the prospect of financial gain. The expeditions of Samuel Hearne, particularly his epic 1770–1771 journey across the barrens to the mouth of the Coppermine River, contributed important geographic knowledge and fixed a point on the Arctic coast far to the west. A decade later in 1789, Alexander Mackenzie reached the Arctic mouth of the river thereafter named for him, and in 1793 he demonstrated just how broad northern North America really is.

Nineteenth-century exploration was spurred by a sense of national pride, preoccupation with science, and the urge to discover the fate of lost countrymen. After the Napoleonic Wars, left with a large, unemployed navy, Britain undertook to lead the European nations in Arctic exploration, and proceeded in a systematic, scientific way to find, define, and map the Northwest Passage. Beginning in 1818, both land and sea expeditions helped to draw large parts of the northern map. By the 1840s the Parry Channel coasts were known, except for those along its western end, and Alaska's coast on the Beaufort Sea had been mapped, as had the Canadian mainland coast as far east as Dease Strait, roughly to where the settlement of Cambridge Bay on Victoria Island stands today. The Royal Navy had amassed much new scientific knowledge about the geophysical and marine environments, and valuable ethnographic accounts of Inuit life had been recorded. Yet the passage itself had not been completed, although the options of where it might lie had been narrowed.

One last expedition was mounted in 1845. Sir John Franklin, an aging veteran of polar navigation, won the command. His two ships, the *Erebus* and the *Terror,* sailed into Lancaster Sound, headed west and northwest around Cornwallis Island, and put up for the

Inuit built these human-like cairns, called inukshuks, *as guideposts in the trackless tundra. Today, the form is widely copied as an emblem of the true north. Photograph by DIAND.*

winter of 1845–1846 at Beechey Island. The next summer, they pushed south through Peel Sound, only to be trapped in the ice in Victoria Strait. We now know that, after Franklin died in 1847, the crew abandoned ship in spring to march south—incidentally, completing the Northwest Passage—toward the previously mapped mouth of the Back River and inland to some fur trading post. Their retreat was never completed and all 129 members of the Franklin expedition perished.

In Britain, anticipation turned to romantic anxiety over the fate of the expedition, and massive efforts were initiated to save the crew and solve the mystery. Searches were carried out primarily by sea, but also over land with sleds hauled by men in springtime travel. During the next decade no fewer than forty expeditions set out; in one season alone fifteen ships and crews plied the Arctic in search of Franklin. Evidence of tragedy came in 1854 when John Rae brought home Eskimo reports and articles from Franklin's ships. Finally in 1859, Leopold Francis M'Clintock found documents left by the ships' crew on their retreat journey that confirmed their intentions and described their fate.

The search for Franklin also became a rush to map as much shoreline as possible. Indeed, most island coasts on either side of the Parry Channel were mapped during the two spring seasons of 1852 and 1854.

The far north of the Arctic was mapped during exploration related to the urge to reach the North Pole. The United States was well-represented in late nineteenth-century expeditions, and the final triumph of Arctic exploration came when Robert Peary reached the pole in 1909. Otto Sverdrup added 30,000 square miles of land to the map by exploring the central Queen Elizabeth Islands between 1898 and 1902. Stefansson found new land north of Banks Island, and the last unknown islands in Foxe Basin were discovered by air photography in the late 1940s.

The impact of exploration upon the Inuit was not great, although during the early exploration period, some were killed in hostilities and others captured and taken to England as curiosities. The sustained contact of the nineteenth century was more important. In some places Inuit and Europeans established continuing contact as ships wintered over, at times for more than one winter. Presumably, the Inuit gained information about the foreign material world by observation. Certainly, a number of abandoned ships and depots were exploited to great advantage for the new materials of metal and wood. Inuit people today still relate the stories of the wonders that the wooden ships held for their ancestors.

WHALING

Whales were killed primarily for the oil that was rendered from their blubber. Whale oil was almost the only source of oil for lubrication and lighting fuel. Many uses were also found for the flexible baleen comb, or 'bone,' from the jaw, especially as the manufacturing era emerged in the nineteenth century.

Arctic whaling continued for centuries. In the 1600s, whaling ships from the Netherlands and Denmark hunted the west coast of Greenland, and from the beginning of the eighteenth century, there is a record of continuous and expanding whaling activity. The Canadian Arctic waters were visited only occasionally before the Royal Navy proved in 1818 that, in the far north, it was safe to cross to the west side of Baffin Bay. After that, whaling ships moved into Lancaster Sound and down Prince Regent Inlet. They cruised south along the Baffin coast to establish shore stations in Cumberland Sound in 1853 and extended the pursuit into Hudson Strait and Hudson Bay. Before 1810 the whalers were from the Netherlands, Denmark, and Germany, but thereafter they were increasingly British. Also, after sporadic earlier trips, American whalers—based principally in New Bedford, Massachusetts, and Mystic, Connecticut—joined in after 1845. Although the number of ships by that time was in decline, the ships were larger and carried more whaleboats and new killing equipment, and the practice of wintering over was commonplace—all of which increased efficiency. Steam power came along after 1857, and in the western Arctic, Americans brought steam-powered whaling ships through the Bering Strait into the Beaufort Sea in 1889. They wintered at Herschel Island and Baillie Islands and conducted a short-lived but vigorous fishery, which collapsed entirely when the prices of whale oil, and particularly baleen, dropped dramatically after 1904. Substitutes were found for these expensive products, and whalers did not return to the Beaufort after 1908 or to Baffin Bay and Hudson Bay after 1915.

Contact between Inuit and whalers occurred as soon as whalers crossed Baffin Bay to the west side; Clyde River Inuit may have been the first encountered in 1820. When the pattern of the whale-hunting grounds and shore stations became established, the economic link between Inuit and whalers settled into a routine. Rendezvous points sprang up on Frobisher Bay, Cumberland Sound, Pond Inlet, Southampton Island, Herschel Island, and in other areas.

Cape Haven Whaling Station in 1903. This photograph, taken by the legendary geologist and explorer A. P. Low, was of an age already past in 1903. Industrial substitutes for baleen and whale oil ended this incursion into Inuit life as quickly as it began. Photograph by A. P. Low, GSC collection, National Archives of Canada, PA 53586.

To be sure, the two societies, living together, learned from a mutual exchange of social and human comforts. Whalers, for example, valued Inuit labor: Inuit men hunted game to feed wintering whaling crews, used their dog teams to transport men and equipment, and proved to be skilled ice navigators and members of whaleboat crews; Inuit women produced skin clothing and sleeping robes. For the most part, Inuit were paid in kind with manufactured items and materials, and rarely with food. But there were many consequences for the Inuit, including the localization of people to certain sites; dependence on trade items; and disruption of the traditional domestic routine to hunt, trap or work—with the effect that Inuit risked success in providing for their own needs. In addition, trying to provide wintering crewmen with meat put heavy pressure on game animals and left the stocks depleted, to the disadvantage of the people left behind. Diseases foreign to the Inuit were introduced, and many populations were badly reduced; a contagion ultimately carried off the entire Sadlermiut population of Southampton Island. And, in the west, the whalers' alcohol brought with it a certain level of debauchery.

Prolonged whaler–Inuit contact was highly localized and did not affect all Inuit. But those natives who took part in this socio-economic exchange learned new values, assumed new technology, and, above all, became dependent on outside goods. As the whaling ships faded over the horizon for the last time early in the twentieth century, their crews left behind some desperate people who were now poorly prepared to exist without that outside contact.

FUR TRADE

Besides working for wages in kind, the Inuit traded furs and other goods with the whalers. The demise of whaling was as hurtful to the ships' crews as for the Inuit, and several whaling men took up fur trading. Some whalers established posts in the north, while others outfitted coasting boats for annual cruises to bring in goods. Between 1911 and 1916, only a few years after whaling had ended, trade posts opened for Inuit in the Baffin, Beaufort, and Hudson Bay areas. Within a decade it became possible for all Inuit to trap and trade furs at nearby posts, and most Inuit added this activity to their usual round of migration. With the trade came the manufactured technology and dependence upon the supplies to keep it operating. White fox and wolf skins were central to the trade, but traders also sought polar bears, sealskins, plus many other species. For a time there was such a profusion of independent fur traders that the competition was locally advantageous to the trappers.

The komatik *may be pulled by a Ski-Doo and the spear or harpoon traded in for a modern rifle, but hunting on the sea ice still requires strength, skill, and a traditional white blind. Photograph by DIAND.*

Gradually, however, the Hudson's Bay Company came to dominate the trade and its pattern of operation led to a rationalizing of outlets. The sites that survived this "downsizing" form the basis for the present settlement pattern in the Canadian Arctic.

MISSIONS

The presence of the Christian church among the Inuit is also associated with whaling. An attempt to introduce the Moravians to the Cumberland Sound whaling station in 1857 did not succeed, but the Anglican missionary E. J. Peck opened a mission there in 1896, and a missionary has been present in Cumberland Sound from that time on.

Peck and his colleagues introduced a form of symbolic writing for the Inuit language that Peck had first developed from Pitman shorthand for use among the Cree Indians. The script uses sixteen symbols, today referred to as 'syllabics,' that represent sounds. With the script, the spoken word can easily be rendered into writing. Peck's orthography is now standard for the Inuktitut language spoken by the Inuit of the central and eastern Canadian Arctic. Peck's converts were first to learn the writing system; a group near Great Whale River (Kuujjuarapik) were taught in 1876, and another in the Baffin region twenty years later. With church literature translated by Peck and printed in syllabics, native lay preachers spread the written word quickly, at times adapting the message to local or personal needs. Inuit literacy, therefore, has existed for over a hundred years.

The Anglican mission also established residential schools for selected Inuit students, and brought in missionaries with medical skills to establish small hospitals. The Roman Catholic church, especially the order of the Oblate Fathers, was equally committed to the Christian mission. Because of these two influences, present-day Arctic settlements tend to be predominantly either Catholic or Anglican. Until 1945 the churches alone provided for Inuit education and health needs. During the early twentieth century, Christianity, and particularly its ministers and priests acted as strongly stabilizing forces in an Inuit emotional and spiritual life that was greatly shaken by the earlier experiences with the whalers and by the uncertain future after the whalers left. The preeminence of the church as a force for social leadership faded with the onslaught of government programs and the administrative infrastructure that accompanied them.

GOVERNMENT

Canada governs its Arctic, of course, but at the beginning it did so from a distance. The Canadian government assumed full responsibility for the territory when the Arctic islands were transferred in 1880. The first on-site government authority arrived with the opening of Royal Canadian Mounted Police posts, at Herschel Island in the west and Fullerton Harbour on Hudson Bay in 1903. The mission of these outposts was to demonstrate the government of Canada's sovereignty by bringing the force of law to the dealings between Inuit and whalers, especially in the west, and to collect customs dues from local trade. The opening of additional police posts in the High Arctic as well as seasonal patrols by government ships reinforced Canadian authority. In addition, a quiet band of government scientists added, season by season, to the outside world's knowledge of resource potential and the natural order of the far north. The concerns for native society were believed to be in the good hands of the churches, which the government subsidized modestly. Besides, the Inuit were renowned as masters of their environment and, it was thought, should be unmolested in their realm.

That was about all the attention the government gave to the Arctic regions until World War II. With the war came the strategic need to provide serviced air routes to Europe and via Alaska to eastern Russia. Major air bases and staging points were built at Fort Chimo (Kuujjuak) and Frobisher Bay (Iqaluit) with the heavy presence of American troops, construction workers, and money. The local Inuit were attracted to these sites, got wage employment, learned some trade skills, salvaged surplus material, and were introduced to the use of cash. Both airbases persisted and eventually became the nuclei of large, lasting settlements in the eastern Arctic.

After the war, the people of southern Canada continued to become more aware of the far north. Transpolar flights and the Cold War necessitated building High Arctic weather stations, bases, and a chain of manned radar stations, the Distant Early Warning (DEW) Line, across the continent at latitude 70° N. DEW Line construction in the mid-1950s was a major but short-lived opportunity for many Inuit to acquire money, skills, and material goods.

When reports of starvation among the Inuit became public and their economic distress better known in the south, civil servants who had long understood the difficult realities of change for northern people gained support for new initiatives. They believed that education, for example, should be as available in the north as it was in the south, for without it Inuit would have no opportunity to participate in modern society. Schools were built at outposts across the north, but, to attend them, students had to live nearby. Soon families gathered round these settlements and stayed. The government responded by establishing housing programs and a health-delivery system. Settlements became permanent and, for the most part, people came in off the land to a new urbanization. They found there new neighbors arrived from outside and mainly in charge.

Everything needed to provide the infrastructure for a small functioning town had, over time, to be put in place. This included roads, electricity, water and sewer systems, airports, telephones, and television, plus all the people and equipment to make them work. It included, too, all the social and political organizations, committees, councils, societies, and associations formed to define the public will and mobilize public action. The standards set in the north were the same as those of small towns elsewhere in Canada, and so the southern way of doing things came to dominate many local and most external matters. That there are today many Inuit who, in their own lifetimes, have experienced the complete spectrum of living—from a confident, self-sufficient, autonomous condition, into a world that is difficult to predict and harder still to influence—is a measure of how dramatic, rapid, and profound change has been in the Arctic over the past four decades.

INUIT: THE PEOPLE

Outsiders have a hard time shaking the conventional image of the Eskimo, or Inuit, as people who must be wonderfully wise and skilled to survive in the Arctic, but who are perhaps too innocent and trusting of the sophisticated white people and their modern ways. The reality is that Inuit are wise enough to assume selectively the modern ways of the whites and that they have matched their new skills against new situations with a sophistication that can only be admired and respected. But they are still Inuit. They are born and brought up in the same natural Arctic environment as their forebears; it is a conditioning environment that requires the same age-old responses to ensure comfort and security. The power of the environment keeps people together in cooperation, and consensus is strong. That the values and beliefs of the traditional culture are in force today is not always evident to the casual observer, but those values are within reach of each Inuk, even by the young people who might have difficulty defining them, but who nevertheless know their own identities. (Note: Inuk is the singular of Inuit.)

About 25,000 Inuit live in Canada; of these, 1,500 reside in Labrador, 5,000 in Northwest Québec, and the remainder in the Northwest Territories. No Inuit live in Yukon. Although no one really knows, it is believed that the level of population is about the same size today as it was at the time of contact with the outside. In the intervening years, many Inuit groups were devastated and, in places, entirely wiped out by diseases introduced by the Europeans. The Inuit were so vulnerable that infections that merely

Inuit Population in Canada, 1986

Legend:
- < 200
- 200 – 500
- 500 – 900
- > 900

Source: *Statistics Canada*, Special Tabulations, DIAND

annoyed the carriers frequently proved lethal to the natives. Whooping cough, measles, diphtheria, and tuberculosis took heavy tolls. Smallpox, syphilis, scarlet fever, and mumps spread on contact. Estimates of the Inuit population were as low as 8,000 to 9,000 in the 1930s.

Today's population is much healthier, and the reward for containing disease is plenty of children. By standards set in the rest of Canada, the Inuit are undergoing a population explosion: About 45 percent of the current population is under fifteen years of age. Fertility rates run very high, and the risks of unattended births, without modern medicine, have greatly diminished, since 90 percent of Inuit babies are now born in hospitals. Most expectant mothers must travel away from their homes to the hospital and leave their families, which has provoked a backlash to this system. A strong movement is now underway to bring back midwives so that Inuit women can give birth at home. In general, health delivery is excellent, with population growth the result.

The death rate for Inuit resembles the death rate for all of Canada. In fact, in the last few years, the Inuit rate has been lower than the national average. Because the base population is so small, slight changes in the number of recorded deaths cause the rate to fluctuate. The rate today is approximately 5 deaths per 1,000 population annually. Causes of death, however, are unlike those in the rest of the country. Accidental death and poisoning are the biggest killers, accounting for about 25 percent of annual deaths. This category includes drowning, freezing to death, and gunshot wounds attributed to misadventure of some sort. The incidence of accidental homicides, fatal fights, and suicides is unsettlingly high. Alcohol is often involved, especially in homicides and freezing deaths. Poisonings may result from improper storage or cooking of meat or from drinking the wrong alcohol. Other causes of death and frequency rates are much the same for the Inuit as for other Canadians. Suicides are a real problem and are distressingly common. In 1989 alone there were 29, and all of them were young people. Whatever the specifics of each incident, this tragedy clearly signifies that social and individual stresses can result from the conflict of cultures and bewilderment in the face of a rapidly changing world.

EDUCATION

If the intrusion of an alien culture is the principal cause of Inuit social malaise, then those who introduced it are now trying to use government-funded education to alter the Arctic social environment and prepare Inuit people for the new age. Federal schools were started in the 1950s. Access to education improved when a system of centrally placed schools with dormitories for students from outlying settlements gave way to providing schools in each community. The Inuit wanted their children educated close to home, and now all settlements have schools that take students through at least grade 9, and most have schools with grades 10 and 11. Completion of high school (grade 12) still requires travel to and living in places such as Iqaluit, Rankin Inlet, Yellowknife, and Inuvik.

A problem remains, after three decades, with keeping children in school through the higher grades. At least half of all Inuit have not completed grade 9, and there is a cohort of people over forty years of age with little or no formal education. Even among those Inuit who have completed grade 12, the number with postsecondary education is still measured in tens rather than hundreds. The failure of many Inuit to stay in school is caused, in part, by a lack of complete understanding on the part of parents of what formal education means and what benefits it offers. Home environments that support study still need developing. Early curricula were imported from southern Canada and were frequently irrelevant to northern lifestyles. Now much of the teaching material incorporates Inuit culture and the Arctic environment. Instruction in English also created a problem. This has changed, too. Children today learn in their own language, Inuktitut, during the first three grades and gradually pick up English. Success rates are rising—children remain longer in school. Boys still tend to leave early, because they want to get out on the land or they have trouble retaining interest in learning things they have only limited opportunity to apply. Girls stay in school longer, and this is reflected in local employment trends, as many young women get office, clerical, or administrative work in their communities. Relatively few people have gone to university, although postsecondary education (away from home) is funded in full by government programs. Some universities have moved north. McGill University in Montréal, for example, conducts a teacher education program in Inuktitut in the Arctic.

Among the Inuit, the most popular kind of education is the acquisition of job-oriented skills. Adult education is available everywhere not only to prepare individuals for wage jobs, but also to teach social skills that enable students to cope with the new ways. Home management and domestic finances, home health procedures, nutrition, first aid and safety, and how to run meetings are examples of subjects offered. Northwest Territories (NWT) Arctic College, with satellite campuses in the larger centers, offers many vocational training courses and is practically oriented for people entering the wage economy. Students receive full financial support for this coursework, and, since many are mature students with family responsibilities, they are usually paid as if on the job. The Arctic College is developing courses in academic subjects so that Inuit students can transfer to other colleges and universities, easing the transition for Inuit students who are willing to pioneer.

It is worth pointing out that each community is anxious to include in the education of its children an understanding of life on the land. Traditionally, this was learned by living on the land, but few now live outside the settlements on a continuing basis. The land-based experience for young people is confined to staying at spring fish camps or accompanying a hunter on occasion. The larger the settlement, the more acute the problem. To make up for the absence of firsthand experience of traditional life on the land, cultural programs have been established in the schools. Elders and others with traditional skills demonstrate them to the children. Girls are shown those things for which women are traditionally responsible, such as sewing and skin preparation. Boys learn how to make tools, hunt, and build an *iglu*. All the children go on a field trip to practice what they have learned. Such programs cannot replace the realities of the old days. They do, however, help Inuit young people to keep in touch with the heritage of the people.

HOUSING

When the schools were built, parents naturally wanted to be with their children. Tent towns and shack houses made from discarded building and packing materials sprang up. Because families occupied one house for a long time instead of moving about, the houses quickly became unhealthy. The need for better housing became obvious.

Beginning in the early 1960s, the government developed a program to prebuild small houses that could be shipped where needed and reassembled on site. The first houses were "five-twelves," so called because they enclosed an area of 512 square feet, and oblong "match-boxes." Designed for the Arctic climate by the Canadian National Research Council, they represented what was thought to be a scientific approach to the housing problem. Little concern, however, was given to social aspects of house design. After all, most Inuit traditionally lived in one room, and so the prefabs came this way. But the new one-room houses for the Inuit strongly contrasted with the two- and three-bedroom houses that the government built for its employees. Inuit easily saw the benefits of more space and a multiple-bedroom configuration, and so the "scientific" design quickly gave way to conventional bungalows and the housing we see today.

The housing program began in earnest after 1964, when 1,300 houses were brought in and distributed all across the North. Houses were to be rented, and a schedule of rents was designed to take into account wages earned. At first, the minimum rent was $5 a month, and the maximum was a quarter of wages paid. Each unit was

The 1987 sealift brings a new house to Grise Fiord, shown against a background of completed models from two earlier years. Photograph by H. Swain.

supplied with basic appliances such as a furnace, stove, refrigerator, and washer and dryer. Permafrost required that water be delivered by truck to indoor tanks and human waste carted off in plastic "honey-bags." Gray water spilled on the ground. Designs improved with time, and in some larger settlements a system of "utilidors," pipelines linking houses, was developed to deliver running water and carry away sewage. To support the program, all the infrastructure for local management, including advisory committees, repair services, and the like, had to be developed and paid for.

Actually, the costs of all these services could never be covered by returns from rent and large government subsidy was necessary. Also, many communities fell chronically behind in their aggregate rent. Shrinking budgets forced some rethinking, and recently a

Home Assistance Program was introduced by which the cost of a house is provided, but the occupant assembles it and takes responsibility for all servicing and maintenance in return for eventual ownership. This program has not appealed to everyone, because many cannot afford the cost.

The housing shortage seems ever-present, in part because family formation is increasing rapidly. Everyone wants his or her own house. Also, older people, who in earlier times would have lived with their children, now have the means to afford houses of their own because they receive Old Age Security and Guaranteed Income Supplement cheques. The housing shortage puts considerable pressure on local housing associations, which are responsible for all the political and advisory functions, including the allocation of units.

The combination of building schools and introducing housing programs has, in effect, urbanized the Arctic. Practically all people now live in permanent settlements. This change from living on the land to living in a modern town has transformed traditional Arctic life; and all the physical and social services that are necessary for urban life have led to a very different pattern of living for Inuit in the far north.

HEALTH

Concerns for the health of the Inuit were always in the minds of those outsiders who knew them well. The missionaries saw the miseries of unattended illness and injury and brought the first doctors into the north to work in the primitive mission hospitals. But these caring people, as well as some civil servants who traveled north, were unable to find resources with which to tackle Inuit health problems, or indeed many other problems among the Inuit, until reports in the 1950s of starvation among the Caribou Eskimos of the Keewatin reached a wider Canadian audience. The Canadian public was stunned to learn that people in Canada were starving. This reaction created the political will among leaders to allocate public funds to improve the situation.

Soon after, the Eastern Arctic patrol ships instituted medical inspections of the Inuit. One result was the discovery of a high incidence of tuberculosis. The government decided to bring the patients to sanitaria in southern Canada, where they arrived in increasing numbers until, in 1956, one Inuit in seven was in hospital. Many died there; some were even buried in the south. Others remained for months or years before they were cured, only to leave the modern comforts of hospital care to return to the family tent or *iglu* on the shores of the Arctic seas. The personal experience was traumatic for both those who were sick in an alien world and for those who stayed behind not knowing the fate of their loved ones. The threat eventually passed, however, and for all intents and purposes, tuberculosis was eliminated. Other diseases persisted, of course, as did a continuing concern for the general health of the people.

When settlements came into being after World War II, the need for on-site health facilities led to the building of nursing stations at each place. The nurses, many of them, originally, adventurous Australian and British women, are trained to cope with a wide range of sicknesses and injuries and frequently are called upon to perform simple medical procedures that elsewhere are reserved for doctors. The stations have one or two hospital beds, usually an X-ray machine, and a comprehensive stock of medicines.

The health care system also provides for medical evacuation of serious cases to hospitals in southern Canada for consultation and treatment with specialists; for example, nearly all babies are delivered in southern hospitals. The system works well except when time is of the essence, since flying takes time that imposes a risk for patients who need emergency treatment. In addition, bad weather can prevent flights, with even more threatening consequences for the ill or injured. Such worries have caused the larger communities to appeal for resident doctors, and settlements with more than 1,000 people are now likely to have a physician. Regional hospitals in Inuvik, Iqaluit, and Yellowknife also serve people within the Territories. Canada has a comprehensive medical insurance system under which the costs of Inuit health care are borne by government budgets. In 1987, payments for the medical services of doctors, dentists, and specialists in the Northwest Territories, which has a population of 53,800, were $10.8 million. All health care costs to cover nursing stations, travel of patients and families, and hospital costs inside and charges outside the Territories, plus the services of doctors, equalled about $150 million.

Inuit health problems are similar to the health problems of any society, although certain conditions are more prevalent among the Inuit. The incidence of ear infections in children is unusually high, probably because the incidence of colds is high—every time a plane comes into a community it brings new bacteria that local residents may not have resistance to. There is less attention paid to casual infections than in the south. Among adults, the noise of rifles, snow machines, and all-terrain vehicles may have negative effects on hearing. Frostbite is more common now than in the past, because snowmobiles move faster than the traditional dog sleds; the higher the speed, the more likely is frostbite.

The considerable changes in diet brought on by modernization have led to nutritional problems. Poor eating habits are evident among the youth; many have taken to "junk food" with the passion of kids everywhere. Junk food diets, together with the wide use of sugar, has affected dental health. Dental decay, once unusual among a people who ate only meat, is now common. To address this problem, Inuit dental technicians are available to fill cavities and perform other common maintenance routines. The new diet, with reduced use of country food, may also be contributing to an increase in cardiovascular problems and may have played a part in the appearance of diabetes, which was previously unknown among the Inuit. Also, the use of tobacco has produced a rise in lung cancer rates. As if to show how integrated the north and south have become, the tragedy of AIDS has claimed an Inuit victim, and more cases are likely to occur.

WORKING FOR WAGES

In today's world, the Inuit cannot exist without cash. There may be those who can drive dogs, feed the family on game, and seem to be self-sufficient, but even ammunition costs money and trappers face a wide range of equipment costs. Almost everyone owns a snowmobile, an all-terrain vehicle, and an outboard motor—for all of which they must buy expensive fuel, in addition to paying for the items themselves and their upkeep. All families now keep house with the aid of soaps, paper, and other household supplies that the general store provides, and every family buys quantities of prepared and fresh foods.

Wage employment is a solution, and the infrastructure of settlements today provides that opportunity through many service jobs. Janitors, repair technicians, truck drivers, skilled tradespeople, straight labor, store clerks, office workers, radio operators, mechanics, and short-order cooks are needed. Small communities may need all these positions, but may require only a few people to fill them. All communities need management of these tasks, however, and for many years senior management jobs were held solely by whites from the south. Education is opening the doors of the management suite to Inuit, however, and the southern administrator is disappearing from the north. More Inuit are assuming the salaried responsibilities at the top, which include recruiting and retaining employees. Scarce jobs are closely held, although there is some turnover, especially at the lower end of the wage scale. Interestingly some of the best and most respected hunters and elders who are serious about family responsibilities are the first to sacrifice their independence and love of the land for 9-to-5 work and a secure income.

The emerging crisis is the increasing number of new entrants to the workforce. The Inuit population explosion is creating a situation that requires the creation of jobs at an increasing rate in an environment where a limited economic base and isolation restrict opportunity. How can 15,000 working-age Inuit—who want jobs and are spread over 2 million square kilometers (770,000 sq mi)

1,000 kilometers (600 mi) from the settled part of the country—be employed? Their distance from the center of social gravity in Canada, plus the fact that there is no real out-migration of Inuit from their homeland, raises the very serious question of how wealth is to be created in the north.

One must not forget, in contemplating these questions, the importance of native hunting and traditional food procurement. These practices create wealth; they add value to the northern balance sheet. The value of hunting is hard to measure, but a realistic estimate is that, in most Inuit communities, more than half of the people gain more than half of their food from hunting and fishing. In some remote and smaller settlements, the dependence upon country food is greater. The trapping economy must also be taken into account. Because all Inuit now live in towns and often are free from wage work only on weekends, trapping has declined. In fact, except for the elders and near-elders, trapping is relegated to overtime work. The wealth generated by trapping, despite volatility in the fur market, seems to be steadily declining. In part this is due to vagaries of fashion, but also to the rising tide of environmental consciousness. Furs have become unpopular with the international public, and the northern trapper is paying the price. The fact that trapping is still a significant part of Inuit income and a source of personal pride and social esteem to the trapper himself indicates that more than economic issues are involved.

Some sources of wealth (i.e., types of work) have not attracted many Inuit so far. Mining is an example. Several mines operate in the Northwest Territories, most of which were opened in the 1980s. All of the mining companies have been sensitive about providing jobs to native northerners. Indeed, training programs, favorable work rotation schemes, and excellent wages are all standard. At first Inuit interest in such jobs ran high, but, over time, employment settled back. Inuit workers make up no more than 20 percent of the current mining workforce. In another generation the rate may climb, but mines run out of ore and mining has always followed a boom-bust cycle.

The other non-renewable natural resources in the north are oil and gas. So far, most activity has been limited to exploration, which is well-suited to Inuit participation, in that it is temporary, high paying, and seasonal. The skills required by the industry to exploit these resources fully, however, have limited native participation to the lower end of the pay scale. If production is ever established, it will not be especially job-intensive.

One new growth industry in the north is the rise of the Inuit civil servant. For so long, Inuit have been passive in their relations with the dominant society. That is no longer the case. The community life of settlements first produced the need for social organization and administration. Inuit are gradually assuming all the positions in the bureaucracies of their communities, and rightly so. They have also formed their own organizations: the Inuvialuit Regional Corporation in the western Arctic; the Tungavik Federation of Nunavut, representing most Inuit in the central and eastern Arctic; and Makivik Corporation in northern Québec. Moreover, there is now an Inuit Circumpolar Conference, a pan-Inuit group with members from Siberia to Greenland that promotes an international agenda. It is not a surprise today to notice a briefcase-carrying Inuk traveling north or south on business.

CARVING AND CRAFTS

No other source of wealth in the north has the distinctiveness of Inuit art. It embodies a keen perception of nature, reflects a superb ability to adapt a medium to depict reality, and expresses an almost spiritual message in what is created. The best of Inuit soapstone carvings, graphic art pieces, and woven or sewn wall hangings are fine artistic creations by universal standards. In addition, many objects that are more craft than art are produced for the souvenir trade.

Inuit art, as we know it through galleries and other outlets, is a recent phenomenon. In 1947 the artist James Houston went to northern Québec as a civil servant. He encouraged the Inuit to draw and carve, and he brought back the results to Ottawa where they received public acclaim. The next year, new production was sold immediately, and Inuit art became highly desirable. Soapstone carvings of animals, people, and scenes from Inuit life were appealing for their subject matter and for the uncanny authenticity and simple beauty of their representations. Carvers branched out into other materials, using the bleached whalebone that dotted the shoreline, or carving in antler bone, musk-ox horn, or ivory. Over time, certain individuals became artists well-known in their own right, recognized on an international scale and respected for a sensitive rendition of the Inuit world.

Part of the uniqueness of Inuit soapstone carving is that different qualities of soapstone became associated with the individual locations where it is quarried. One can distinguish Baker Lake stone carvings from Lake Harbour stone carvings, not just by the artists, but also by the look of the material. Problems with quarrying and finding new sources of stone have resulted in shipping stone from alien locations all over the Arctic, which destroys the distinctiveness of associating a certain stone with each location. Some stone has even been imported from down south.

Houston also introduced and encouraged the graphic arts. Today at least five or six communities are renowned for prints or screened products. Cape Dorset is best known, but artisans in Baker Lake, Pangnirtung, Povungnituk, and Clyde River have also produced works of high quality. Early on in the print business, the Eskimo Art Council was established as a panel of individuals knowledgeable in the arts to act as a jury for submitted prints. The annual production from an individual print shop had to be approved by the council to be released to the market. Formation of the council proved to be a wise step in quality control: Inuit prints have maintained a high value as a result.

A further development of Inuit artistry is also represented in wall hangings, tableaux of Inuit life created by sewing cut-out figures onto a cloth background. All of the wall hangings have a certain charm, and some have the qualities of genuine art. The wall hangings, as well as the drawings and prints, of Jessie Oonark from Baker Lake, for example, evoke a sense of awe and appreciation.

Unfortunately, not everything that is produced in the north deserves to be called art. When soapstone carving first began, everyone in the Inuit communities was encouraged to try it. Now that some handsome prices are paid for the work, anyone who wishes may pose as an artist. Much carving activity goes on in response to the need for cash, perhaps more than is the result of artistic discipline and inspiration. Just before Christmas, for example, it seems that everyone is at it. The result is a lot of pretty ordinary, and even bad, products.

The problem of quality control is compounded by the buying system. The Hudson's Bay Company (now called Northern Stores) and other private buyers are able to exercise choice, buying what they like, need, or know will sell. The other outlets to the south for crafts and art are the cooperative federations. The co-op stores have a central warehouse in the south to which individual stores ship the carvings they purchase, and a central agency markets them with suitable mark-up. Quality control, therefore, is at the purchasing end, and varies with each manager. Unfortunately, most buyers are store managers, not art connoisseurs. Besides, the carvers who bring in their works are members of the co-ops—the co-op stores belong to them and the manager works for them. Accordingly, managers find it difficult to refuse to buy any member's carvings. As a result of this system, co-ops have paid millions of dollars for inventories of carvings that no one wants.

Over time, it became apparent that certain people really were artists, and the federal government adopted the policy of publicizing the reputations of individual artists through brochures, juried purchases, and support for gallery presentations. The art world came to know the person and the distinctiveness of his or her art. There are now many famous Inuit artists.

The subjects dealt with in carvings have evolved from some of the simplest representations of Inuit life (mother and child, animals and birds, and the hunter) to action pieces (drum dancers, men killing seals or bears, people paddling *umiaks*) to the supernatural (with pieces representing spirits of various sorts). Another development is that carvings are also being made larger and larger; corporate boardrooms and reception centers will buy large pieces and pay more money for them. Some carvers conclude that they will get more cash if they carve larger pieces, but quite a few very large carvings remain unsold.

Carving and print production is now an important part of Inuit life and their image of themselves as a people. That this art form has found favor worldwide has certainly strengthened Inuit self-respect. The economic impact is no less important. In 1981–1982 the average family income from carving was as high as $10,000 in one settlement, although it averaged about $2,000 to $5,000 per family in most carving communities. A current estimate of total payment to carvers is of the order of $10,000,000.

GOVERNMENT TODAY

Government is the biggest enterprise in today's Arctic, and federal departments—notably Defence, Transport, Energy Mines and Resources, Environment, Fisheries and Oceans, and Indian Affairs and Northern Development—actively pursue their mandates. The mandate of the last-named department uniquely calls for its continued reduction in size as its functions devolve to the elected territorial government.

The territorial government has been growing rapidly. By 1980, locally responsible government was a reality, exercised by voters who elected twenty-four members to the Assembly in Yellowknife. So far, party politics has played an insignificant role, with the rather diverse collection of elected members—representing Inuit, Dene, Métis, and white ethnic backgrounds—preferring a consensual mode of decision-making that reflects the traditional mores of the northern peoples. Regionalism tends to be more important than party affiliation.

The territorial government is eager, in principle, to accept devolution from Ottawa of province-like powers and to boost its prestige and legitimacy by seeking parity with the provincial governments at the uniquely Canadian bunfests called Federal-

Provincial (or "First Ministers") Conferences. Ottawa, for its part, is eager to offer up powers. The arguments tend to center around "cherry picking"—the GNWT (Government of the Northwest Territories) seeking to take over only the popular services—and costs. Who will bear the costs is a major, perennial battle. It is estimated that, in 1991–1992, the GNWT will spend $1,077 million, only $175.6 million of which will be raised by taxing its own citizens. Ottawa will provide a block grant of $797.9 million, plus an additional $100 million in other specific transfers.

On a provincial (territorial) economic account basis, in 1989 direct government expenditures of a provincial or local character in the Northwest Territories added up to $15,929 per capita. Add to that the major federal transfer payments to businesses and individuals ($1,580 per capita in pensions, family allowances, unemployment insurance, and the like), plus the current northern expenditures of federal departments ($5,371), and it becomes clear that the rest of Canada pays handsomely to support its Arctic hinterland. The total tax take from all sources to all levels of government amounts to only $9,386 per capita, leaving a deficit of $13,494 for each of the 53,800 people in the territory.

While the GNWT is trying to improve its legitimacy to outsiders, it is under attack from below. Subregional native peoples' political organizations erode the centrality of the GNWT and the authority of its hamlet and municipal councils. Moreover, the GNWT has been forced by the momentum of land claims to support its own demise, through the creation of a third new territory, Nunavut, which would essentially be the homeland of the Inuit. Duplicating functions on that scale will be very costly and will postpone any hopes of true provincehood.

A WALK THROUGH TOWN

To the first-time visitor, Arctic settlements present a landscape that can be easily observed but not always explained. The houses, for example, have a remarkable sameness about them. Made of wood with corrugated metal roofs, they have triple-glazed windows and often a front porch. All Arctic houses are raised off the ground, either on gravel pads and wooden blocks or on piles sunk into permafrost. The houses look alike because all have been provided by the government; often the parts were prefabricated, shipped, and then assembled on site to standard specifications. Efforts are made to vary the design, but with limited success. Most houses are rented. The rent itself is based on the tenants' incomes and bears little or no relationship to the costs of heat, light, water delivery, and waste removal, all of which are provided with the houses. The actual costs of these services are much higher than rental revenue; subsidized housing is part of northern life.

The front porch is a typical feature of Arctic life, a place to protect equipment from the snow and to store frozen meat. It also serves as a buffer that insulates the inside from the cold. The raising of all buildings off the ground is an engineering response to permafrost. To preserve the integrity of ice-rich permafrost, the ground surface must maintain contact with the cold outside air, since a building placed directly upon the earth would destroy its thermal equilibrium, causing the permafrost to melt. Thus, almost all Arctic structures are raised to let the cold air of winter flow freely over the ground surface.

Everyone notices the forest of telephone poles, festooned with electrical and telephone connections. They are especially obvious because there is nothing higher. The poles are anchored in metal cylinders filled with gravel or rocks—another response to permafrost. Tying all the buildings together is a labyrinth of roads built on berms of gravel laid on top of the natural surface, yet another technique of permafrost engineering. Between the houses is a rough surface of undisturbed rock, mud, and gravel, frequently strewn with discarded chunks of building material, cardboard, corrugated metal, and litter. In winter, a cleansing blanket of snow hides it all from view. Community clean-up is practiced more diligently in some places than in others.

Another feature of settlement life that becomes apparent is the native preoccupation with that icon of the industrial world, the

A load of soapstone boulders at Grise Fiord awaits the carver's hands. In the background are other typical sights in a modern settlement: from left to right, generator buildings, equipment shop, fuel tanks, house, playground, school, telephone pole in artificial hole, visiting officials. Photograph by H. Swain.

machine. Machines are everywhere. One, two, or more snowmobiles wait at every doorstep. Those that no longer function are simply moved aside and pirated for spare parts. In summer, people move about in all-terrain vehicles (ATVs), three- and four-wheeled Hondas with balloon tires, that travel anywhere on or off the roads. Down by the shore are the boats—canoes and larger whaleboats with half cabins, all driven by "kickers" (outboard motors). Over by the hamlet garage one can see one or two bulldozers, an earth scraper, gravel trucks, an earth mover, pickups, and vans. All this equipment is designed for heavy-duty roadwork, and it seems strangely out of scale with the needs of a very small settlement.

Then there are the large oil tanks, big enough to store an entire year's supply of fuel oil and gasoline. Sealift comes only once a year. In fact, many aspects of life revolve about the once-a-year sealift and what it brings: bulk orders for staples, new vehicles, prefabricated houses, all manner of heavy equipment. "Ship time" is a high point of the year.

It is easy to identify the houses, painted rust-red, yellow, brown, blue, and green. The other larger buildings clearly have institutional functions. There is the three-bayed town garage, the two-bayed fire hall, the large nursing station, the store, and the municipal office block. The largest and sometimes the strangest-looking building in town, with a play-station outside, is the school. Many of those built in the 1970s and 1980s are very well equipped. From the outside, they look to be designed with pure architectural pretense, and, inside, the spaces have unorthodox shapes that may or may not have to do with imagined environments for learning. Schools with gyms are often the center for community functions. Then, as one walks by, the heavy drone of diesel engines indentifies the local power supply. Somewhere on a hill or at the edge of town stand two or three large, white satellite dishes providing the most sophisticated communication links to the rest of the country. Now television is as present in every home in the Arctic as it is elsewhere, and direct-dial telephone and facsimile transmission makes people around the world seem as close as the house next door. Although information moves over distances with ease, Arctic settlements are still isolated when it comes to moving people, and that means there is risk to a person's well-being when specialized medical help is hours away by airplane.

For all the modern infrastructure of the overbuilt urban setting in small northern settlements, there are still the reminders that an Inuit society occupies them. In many places women and girls still use the traditional outer garment, called the *amauti,* to carry their children or baby siblings on their backs. Outside of some houses sits the three- to four-meter (10- to 13-ft) *komatik,* a traditional sled, with its two long runners tied to cross slats. Somewhere near each town are a few dog teams, six to eight dogs staked out on long chains that keep them separated so they cannot fight. There they

Towns such as Pond Inlet, built on permafrost, allow cold air to circulate between buildings and ground and arrange for trucked delivery of fresh water. The water tank in the house is above the fixtures and is but half the size of the sewage tank below, which is also emptied by truck. Photograph by H. Swain.

watch and sleep until the short summer is over and the snow and their working life returns.

If you visit the Northern Store or the Co-op, note the price of food. Although most staples arrive by sea, all perishables are flown in. This system has the effect of limiting selection and making the merchandise quite expensive by southern standards. Recent surveys, for example, show that the costs of a standard "food basket," when compared with a bench-mark southern city value of 100, are as follows: Iqaluit, 141; Pond Inlet, 164; and Resolute, 179. Bear in mind, though, that Inuit families consume a significant amount of country food, which, when available, they prefer.

CANADIANS AND THE NORTH

Although the north is not an everyday preoccupation with ordinary Canadians, they respond approvingly when reminded that they are a northern people. There are many things in life that reinforce this description, winter temperatures and snow not the least among them. It takes only a little thought to understand the climatic consequences of a northern location—limits to agriculture, high energy costs for heating and working, and special construction techniques and design that must take into account a cold, snowy environment. In fact, life in the north just plain costs more than in more temperate parts of the continent, such as New Hampshire, Wisconsin, and Montana. Nevertheless, accommodations to climate have long been made, and for that vast majority of the Canadian citizenry who live within 300 km (200 mi) of the U.S. (Lower 48) border, life on a day-to-day basis has much in common with the neighbors to the south.

But Canadians are not Americans. Canadians have a different history with a relatively orderly or even stately march of occupation and settlement, in contrast to the pervasive frontier culture of the United States. There has been, in Canada, the abiding and frequently guiding hand of government to help create an orderly sort of progress and to overcome some of the obstacles of nature which might not have yielded without it. Canadians, unlike Americans, have not populated all or even most of their territory. Over their shoulders, looking north, is an almost empty land that stretches to within 680 km (420 mi) of the North Pole. Most people fail to realize the size of this empty land, or the details of what it contains, or what it's really like to live up there. They expect that it has substantial resource potential and that development will occur eventually, and that they might even like to visit the Arctic. What is important to almost everyone is that it is the Canadian Arctic, and it belongs to Canada. Canada is a different country because of this differing view of geography; and it is especially valuable to understand Canadian geography and culture when looking over the other shoulder at the large and vigorous neighbor to the south.

In recent years, Canadians have begun to realize that, along with the pride and potential of the north, there are obligations and costs. They have marveled at the indigenous people, but now the native northerners are telling the rest of the country that Canada must share the good life and accept partnership with Inuit and Dene on a more equal footing. The land claims negotiations are evidence of it. As well, increasing numbers of Canadians have visited the north or have a relative or friend with northern experience. More and more money and resources are invested in the north, and modest in-migration is steadily occurring. The new northerners frequently remind the south of their need for living standards equal to those in the rest of the country. In addition, a frighteningly expensive infrastructure is required to guard the outposts. The distances are vast, but modern technology is applied to the transportation of goods and people and to communications, mainly at government expense. Canadian taxpayers support territorial government operations to the tune of $1 billion annually. This figure does not include private investment, the northern activities of the national government, or personal transfer payments of a universal kind.

A new concern is also pervading Canadian society—for the natural environment. This concern accompanies awareness of the dangers presented by global warming and ozone depletion in the atmosphere. News reports frequently cite the Arctic as an area likely to be affected by these trends most dramatically. Canadians are all the more alarmed to learn that their circumpolar neighbors and the industrial zones of southern Canada are the sources of trouble. These developments have caused the growth of a protective attitude toward Canada's Arctic and an increasing willingness to take—and pay for—corrective steps. Environmental concern has also led Canada into supporting cooperative scientific work on an international level. Canadians, in general, may have an imperfect understanding of these subjects, but they are possessive about their Arctic and regard it as one of the ingredients that gives personality to the nation.

SOVEREIGNTY AND THE ARCTIC

Several times in the last few years a general concern has arisen among Canadians about the status and control of the Northwest Passage. There is no international agreement about the rights of marine passage through this waterway.

The ownership of the islands themselves is not now in dispute. After Otto Sverdrup completed his explorations in the High Arctic in 1902, he persuaded the Norwegian government to press a claim by right of discovery. Canada and Norway carried on a dispute over sovereignty until 1930, when Canada provided an annuity for Sverdrup in the amount of $42,000, and the explorer lodged all of his records with Canadian authorities. Also in the early part of the twentieth century, the right of discovery for Ellesmere Island was raised on behalf of the United States by Robert E. Peary when he was preparing his assault on the North Pole. Various other U.S. expeditions sought to explore the north without authority from Canada, and a U.S. mining operation was even mooted. Canada responded by instituting the annual Eastern Arctic Patrol by government ship, from which several flag-planting expeditions touched many islands, in effect claiming the entire archipelago for Canada. The Royal Canadian Mounted Police opened a series of posts in the High Arctic in the 1920s, and overland police patrols by dog team made an effort to at least set foot on as much territory as possible, thereby securing the Canadian claim.

During World War II, several defense projects in the north were undertaken jointly with the United States. The larger partner exercised its action plan with such independence that concerns—albeit unspoken—arose among Canadians for Canada's authority over its own northern regions. No official action was taken, however, and cooperation prevailed. After the war, Canada assumed full responsibility for the continued function of all joint installations.

Regarding rights to Arctic waters, only the United States refuses to recognize Canada's sovereign control. A challenge was perceived when the *Manhattan,* a tanker whose hull was reinforced for Arctic travel, proposed an east-west traverse of the Northwest Passage via the Parry Channel and Prince of Wales Strait in 1969. Canada maintained its dignity by accompanying the *Manhattan* with icebreaker escort vessels. In response to this event, Canada passed legislation, the Arctic Waters Pollution Prevention Act, in 1972, which established standards for vessels and shipping and general regulations for the prevention of pollution—in particular, oil spills—and which declared a 100-mile (162-km) zone from its shores within which these standards and regulations would be enforced. The act was unilateral and declared not subject to decisions of the International Court. A subsequent passage of the *Manhattan* in both directions through the passage was made with Canadian observers on board as well as icebreaker support. The companies testing tanker transportation later decided not to pursue their research further.

In 1985, the Coast Guard ship *Polar Sea,* without requesting Canadian permission, traversed the passage ostensibly to move the ship from the East Coast to the Pacific. This move caused much comment and concern in Canada, and discussions followed between the governments of Canada and the United States in which Canada sought U.S. recognition of Canadian sovereignty over the passage. Since then, the *Polar Sea* has made three more transits without requesting official permission. The current state of affairs is that the United States has agreed to obtain prior Canadian consent for transit of U.S. government surface ships through waters of the Canadian archipelago, including the Northwest Passage. This is without prejudice to the official position of each country regarding the passage. Canada has also taken steps to protect its legal status by defining the straight baselines from landpoint to landpoint, from which the 12-mile (19-km) limit is measured, by installing hydrophones to detect submarines and by patrolling the passage with military aircraft—bought, of course, from the United States.

LAND CLAIMS

Canada's internationally unique policy regarding native land claims stems from English common law and from enlightened colonial policy at the time of the Seven Years' War (1756–1763). The national government, through the Department of Indian Affairs and Northern Development, today diligently pursues a policy similar to one first enunciated in a royal proclamation in 1763, which stated, in brief, that agents of the Crown should treat with the possessors of aboriginal rights to land before opening the land to European settlement.

Those common law rights were thought to be, then and now, less than fee simple title or international sovereignty (although the closer parity to military and technological might in the eighteenth century might have made "nation-to-nation" agreements more natural then). Rather, "Indian title" was conceived of as usufructuary, heritable, or transferrable only collectively, and as limited to rights to continue such traditional activities as hunting, trapping, fishing, and using certain resources for sacred or social purposes—so long as those rights had been exercised over the given territory exclusively and for a very long time. The essence of the English legal view of aboriginal title was that it was an ill-defined, place-specific cloud on the Crown's right to grant unencumbered land to settlers. It follows that the essential goal of the treaties negotiated since, especially in the Canadian West and North, has been to exchange these vague rights for better-defined ones. That the Indians should feel ambivalent about the results of the survival of this concept of their rights from the Age of Enlightenment is not wholly surprising, but it is another story.

Modern federal policy on land claims was established in 1973 and has been refined several times since then. In essence, claims based on unextinguished aboriginal rights are considered comprehensive claims; any settlements reached are called treaties and are therefore constitutionally protected agreements. Typically, new treaties exchange aboriginal title for defined rights over land and resources, such as specific harvesting and management rights to wildlife, fish, and forest resources. Contemporary agreements also call for substantial payments to the aboriginal parties. Since 1986, aboriginal rights may be retained on lands the aboriginals will hold, and any rights not related to land and resources need not be affected. These rights and other entitlements are in addition to the civil rights of all Canadians.

Modern treaties under this policy have been concluded in James Bay and northern Québec (Cree and Inuit, 1975), in northeastern Québec (Naskapi, 1978), and with the Inuvialuit of the lower Mackenzie River valley (1984). The Inuvialuit Final Agreement, affecting the 2,800 Inuit of the western Arctic, took from 1977 until 1984 before the ratified claim was made law by Parliament. This law provided the Inuvialuit title to 91,000 square kilometers (35,000 sq mi) of land and 13,000 square kilometers (5,000 sq mi) of subsurface rights. Further, the Inuvialuit received a statutory role in wildlife management, land-use planning, and environmental management in the region. They also received cash compensation of $45 million. In return, aboriginal rights were extinguished. Elsewhere in western Arctic, good progress is being made with the Council for Yukon Indians and with the Gwich'in people of the Mackenzie Valley.

In the eastern Arctic and High Arctic, negotiations with the Inuit organization, the Tungavik Federation of Nunavut, have been underway for some time. An agreement-in-principle was signed in Igloolik in 1990, and steady progress is being made toward a final agreement in 1992. The stakes are large. The area in question is all of the Northwest Territories east and north of a line that roughly demarcates the boundary between Inuit and Dene occupancy. The agreement contains complex measures to deal with land tenure, wildlife and environmental management, and even sea ice. It confirms Inuit ownership of 350,000 square kilometers (135,000 sq mi) of land, including 36,257 square kilometers (14,000 sq mi) of subsurface rights. It also calls for payment of $580 million over fifteen years by Canada to an estimated 17,000 beneficiaries and the eventual creation of a third, new northern territory, Nunavut—the first such jurisdiction with a majority native population.

PART TWO

The Itinerary

Canada North

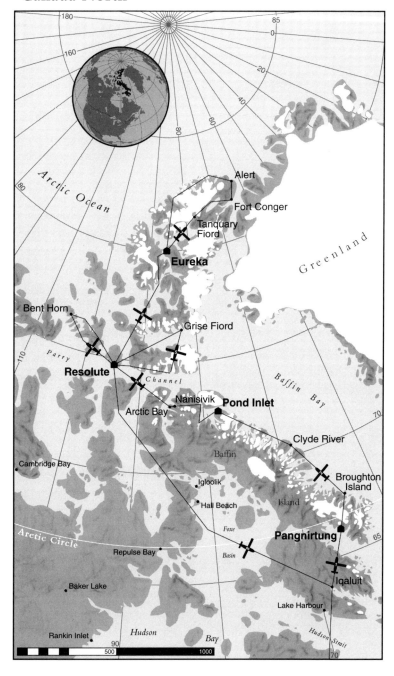

PROLOGUE

Our purpose in this itinerary is to give you as much of a sense of Canada's High Arctic as possible in a short time. Days rather than dollars are the limiting factor in the design of this field trip. Still, there is really no alternative to the use of a chartered short-takeoff-and-landing (STOL) aircraft in this roadless region. Even a regular small plane would mean missing many fascinating places such as Beechey, Ward Hunt, or Prince Leopold islands. The narwhals of Koluktoo Bay, the fossil forest of Axel Heiberg, and the wonderful wildlife of Polar Bear Pass are likewise inaccessible to any but STOL planes or their more expensive cousin, the helicopter.

Once equipped, the North is your oyster. If the weather cooperates—and all plans depend utterly on it—you will see a wealth of landscapes, bare-naked geology, flora and fauna, and ice in a dozen forms. You will see every type of settlement that exists in the Arctic: Inuit, administrative, logistical, mining, scientific, military, expeditionary, and even ghost. Perhaps we should stress the ghost settlement, for since the times of the Tunnit onward, occupation of the Arctic has been contingent and temporary, threatened by the ferocity of the local climate and, lately, by the vicissitudes of distant markets and fashions. What you will not see are trees more than three inches tall, or sunsets during the summer.

In this harsh and sparsely populated region, every single human being takes on a unique character. In fact, it often seems that the region is populated solely by characters, most of whom seem to know each other despite the vast distances, and each can tell the most colorful stories about all the others. All the phases of human history coexist in today's Arctic. There are still people who live mainly on the land (which includes sea ice), there are people who are electronically plugged into the whole global village, and sometimes they are the same people.

The Arctic has attracted all sorts of outsiders, and since the nineteenth century far too many of them have felt compelled to record their experiences in books and memoirs. The vastness of the literature—surely no other region has produced more printed pages per visitor—is a measure of the hold this fascinating region has always exercised on European and American minds. But we are hopeful that *Canada North* meets your approval, for it is written in the spirit of getting the facts right, at the same time we convey some of the mysteries of its cultural and physical landscapes.

⚠ Day One

IQALUIT AND PANGNIRTUNG

Iqaluit

Iqaluit, meaning "fish camp" in Inuktitut, is a traditional gathering place for Inuit. In August of 1576, Sir Martin Frobisher, searching for a commercial sea route to the Far East, entered the bay that now bears his name. The appearance of the Inuit faces he saw convinced Frobisher he was close to Cathay. Five of his men went ashore in a boat and vanished. In return, he captured an Inuk man, woman, and child, with their kayak and hunting gear, and took them back to England, where they soon died. He also collected specimens of plants and mineralized rocks that he believed to be gold ore. The following year, three ships returned to "Frobisher's Streights" to prospect and recover better ore samples. They returned home with 200 tons. In 1578 Frobisher led a colonizing expedition of fifteen ships and 100 settlers intending to establish a mining community, but storms west of Greenland and wind and heavy ice in Frobisher Bay sank several of the ships. The expedition brought 1,000 tons of ore back to England, only to discover that assays showed that the ore was not gold, and the whole cargo was worthless. The traces of this mining experiment are visible today on Kodlunarn Island.

European whalers made regular visits to Frobisher Bay from early in the nineteenth century. They introduced new materials, technology, and customs to the local Inuit, who in return supplied labor, domestic skills and comforts, meat, furs, handicrafts, and

Ottawa to Pangnirtung

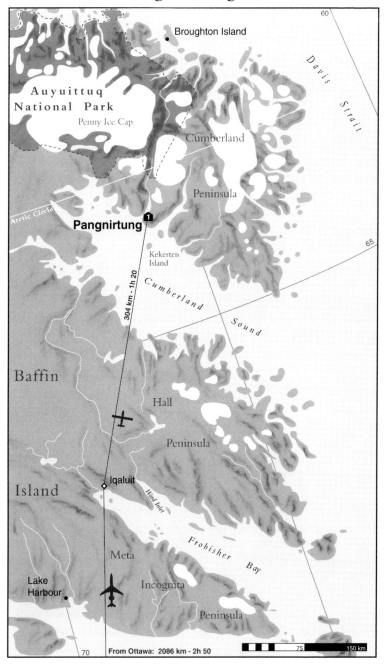

curiosities. In 1914 the Hudson's Bay Company opened a fur trading post at the head of the bay on Ward Inlet. By that time the whaling era had ended, and fur trapping had become the means for Inuit to obtain guns, ammunition, cloth, tobacco, and flour, all of which by then had become necessities. For the next three decades, the hunting and trapping life of the Inuit was uninterrupted. A major impact on the people occurred in 1942 when a military airfield was built right at the head of the bay by the United States Air Force. The field operated as part of the Northwest Staging Route, the World War II path to Europe. After the war, Canada assumed ownership and control of the base, and the Hudson's Bay Company shifted its post to the newly emerging townsite near the airport, leaving a well-preserved relic on the beach in what is now suburban Apex. Strategic and defense interest in the Arctic persisted, and Frobisher Bay, as the settlement became known, grew to serve as a central staging point for the construction of radar defenses and for aerial surveillance operations that protect continental North America from transpolar intrusion.

Native people were drawn to Frobisher Bay, attracted by the chance to work for wages and by the lifestyle of the American visitors. When the Americans withdrew, the Canadian federal government moved to fill the void by establishing its eastern Arctic regional headquarters in the settlement in 1959. Both the Anglican and Roman Catholic churches opened missions, and a federal school was built in 1955. The hospital was completed in 1964, and the Royal Canadian Mounted Police based their regional administration in Frobisher Bay. The government of the Northwest Territories, which now provides most of the government services offered in the far north, made the town its headquarters for the Baffin Region. The name Frobisher Bay was changed to Iqaluit in 1987.

The terrain around Iqaluit is rocky and rough. The landscape shows a perceptible grain, with ridge and valley features—longitudinal islands and island chain—that tend to run northwest to southeast. It is in part the glacial scour that gives this orientation, and it has also exposed bedrock over much of the surface. To the north and east of the settlement are hills, some of them moraines, with relative reliefs of 120 to 250 m (400 to 820 ft). Twenty to 30 km

(12 to 20 mi) away on the other side of Frobisher Bay, ridges rise to over 600 m (2,000 ft). The countryside around Iqaluit is dotted with hundreds of lakes overflowing into one another in a somewhat rectilinear pattern of immature drainage. The coast of the bay itself is very ragged with many rocky islands, some of which are linked by gravel flats at low tide. At the head of the bay the average spring tide is over 13 m (43 ft), one of the highest tidal ranges in the world. Supply ships calling at the port of Iqaluit rest on the sea bottom at low tide, and trucks drive over the exposed flat to unload the ships. The timing of the tide dictates the workday when ships are in port.

July has a mean high temperature of 11.4°C (52.5°F) and a low of 3.7°C (38.7°F); February, the coldest month, has an average high of –21.5°C (–6.7°F) and a low of –30.3°C (–22.5°F). The average annual precipitation is 433 mm (17 in) comprising 192 mm (8 in) of rain and 255 cm (100 in) of snow.

About 3,200 people live in Iqaluit. This number includes residents of Apex Hill, three miles away. A majority, 62 percent, are Inuit. The relatively high proportion of non-Inuit reflects the town's regional importance as an administrative center for the eastern Arctic. Most government agencies maintain offices or regional headquarters here, and southern businesses that work with the government or otherwise serve the Baffin region operate branch offices in Iqaluit.

Demographic differences between the Inuit and non-Inuit populations are best reflected in the labor force. Only 34 percent of the Inuit are in the labor force and, of these, 25 percent are unemployed; 68 percent of the non-Inuit are in the labor force, but only 3 percent are unemployed. The lower labor force participation by the Inuit is understandable because 40 percent of them are younger than working age, and a few are older. Nevertheless, four of every ten working-age Inuit are not in the labor force. Some choose to be independent and not work for wages, but others who might want jobs are culturally unprepared to migrate out of the Territories to seek work. The non-Inuit population, on the other hand, includes few children and no senior citizens. These in-migrants remain in the north only as long as they are employed.

Iqaluit has about three times the population of other big Arctic settlements in large part because it has attracted Inuit and their relatives from other areas. Its size and diversity act as a magnet that continues to cause growth over and above a natural increase. Many visitors travel to Iqaluit on government or private business; others come to use the health or education facilities, to visit family members, or to be tourists. Approximately 70,000 paying passengers arrived in or departed from Iqaluit's airport in 1990.

A walk or taxi ride through town illustrates a level of sophistication in infrastructure and services that might not be expected in a settlement of 3,000 residents. Iqaluit has nearly 200 hotel rooms, as well as several lounges, dining rooms, and coffeeshops. A large department store compares well with those found in many small towns in the south; there are other necessaries such as grocery stores, a bank, a gas station, taxis, a pharmacy, an optician, a photo store, an art sales outlet, a florist, and a travel agency. Service firms include car rental businesses, laundries, a barbershop, a janitorial service, and a translation service. Accountants, lawyers, management consultants, and an architect-engineer also live and work in Iqaluit. There is a weekly newspaper, *Nunatsiaq News*, and CBC radio and television plus four cable channels are broadcast in town. The Inuit Broadcasting Corporation television production studio in Iqaluit has a satellite up-link that enables it to serve all Inuit-speakers in northern Canada. Iqaluit has a recreation hall, an arena, a year-round pool, curling rink (with artificial ice), parks, a playground, and a community center. There are also a library and a museum.

There are three schools in Iqaluit, two junior level and one senior level. More than 800 students are enrolled, including some from other communities who are in residence finishing the senior grades not offered at home. The Iqaluit campus of Arctic College is committed to adult education, technical and vocational training, and academic transfer programs that enable students to transfer for university degree completion. A regional hospital with thirty-four beds and a Public Health Unit look after local health care needs and provide medical support for patients transferred from smaller and larger facilities as part of the integrated system that links

Arctic clinics to southern hospitals. Government-provided social services include a corrections center, a group home for children, and a drug and alcohol rehabilitation facility. Community-based organizations staffed by volunteers assist in special projects that deal with social dysfunction. The Royal Canadian Mounted Police provide police services, and justice is dispensed by justices of the peace and a circuit court.

Despite the dominance of the wage economy—government and service employment, and private sector—local resources are still harvested as part of one's livelihood. People fish for char; hunt seals, beluga, and narwhal; and take caribou and polar bears, in accordance with a quota system. A few people trap white fox and other animals for fur. More residents are involved in stone and bone carving and craft- and garment-making, which are encouraged by the local Inuit cooperative. Tourist packages for fishing, hunting, and dogsled and kayak travel are increasingly attractive to visitors, and provide local employment and revenue.

Iqaluit to Pangnirtung

DISTANCE: 304 km (189 mi)
FLIGHT TIME: 1 hour, 20 minutes

The route crosses the Hall Peninsula and Cumberland Sound. The peninsula is composed of Precambrian rock, and the Frobisher Bay side displays a noticeable grain on the surface, with rocky ridges paralleling the length of the peninsula. This alignment of ridges, caused mainly by glacial scour, affects the shapes and orientation of the rivers and lakes that lie between them. Toward the center of the peninsula, the ground is more evenly rounded, moraine-covered and better vegetated. Note the down-slope streaming of patterned ground, the effect of solifluction that is common in areas of permanent frost. On the occasional soil and turf patches it is possible to see tundra polygons, the "beehive" network of cracks at the surface. Toward Cumberland Sound, the rock spine returns, and high ridges and knobs of pinkish granite appear. The sound is fringed by many bare rock islands, all smoothed by ice. Pangnirtung is reached by flying up the valley of the fiord to the airstrip above the hamlet.

Pangnirtung

Pangnirtung is at the mouth of the Duval River on the south side of spectacularly beautiful Pangnirtung Fiord. The fiord, running relatively straight in a southwest to northeast direction, is about 3 km (2 mi) across and is flanked by the typical trough walls rising steeply to between 400 and 900 m (1,300 and 3,000 ft) on either side. A distinctive hill behind the settlement rises 700 m (2,297 ft). Alpine glacial landforms with truncated spurs and hanging valleys testify to heavy glaciation; indeed, at one time the whole of Baffin Island was covered by ice. Glacial retreat and sea level changes have produced fragmentary raised beaches up to 60 m (200 ft) above sea level. One of these is the flat upon which Pangnirtung stands. The Penny Icecap, north of the hamlet and within Auyuittuq National Park, is a remnant of the glacial epoch.

Weather observations at Pangnirtung reveal a mean high temperature for July of 11.1°C (52°F) and a mean low of 3.7°C (38.7°F). January has a mean temperature of –25.6°C (–14.1°F). The annual precipitation averages 342 mm (13 in) with 161.5 mm (6 in) of rain and 18.3 cm (7 in) of snow.

The home territory of the Inuit who now live at Pangnirtung encompasses the whole of Cumberland Sound and the surrounding inland areas as far west as Nettilling Lake, as well as the land around the Cumberland Peninsula and Davis Strait. They call the sound *Tinikjuakvik,* which means "place of big running out," referring to the strong tidal outflow.

European exploration brought John Davis in 1585 and 1587 and William Baffin in 1616, who both contributed to mapping the area, although their Baffin maps were regarded as fantastic and were discounted for two centuries. John Ross restored confidence in Baffin by confirming his discoveries in 1818. Europeans largely bypassed Cumberland Sound until 1839 when William Penny surveyed the whale resources of the sound and recommended a shore station be established for harvesting bowheads. After 1853, whalers from Scotland set up stations at Kekerten Island just south of the mouth of Pangnirtung Fiord and at Blacklead Island across the sound on the south side, midway along its length. American whalers set up their station at Cape Haven at the southern tip of Hall Peninsula, which divides Frobisher Bay from Cumberland Sound. An Anglican mission was opened at Blacklead in 1894, and the Hudson's Bay Company set up a post at Pangnirtung in 1921, where the Anglican mission was transferred in 1926. The Anglicans built a small school and a hospital that served south Baffin until the regional hospital was opened in Iqaluit. The RCMP came in 1923 and a federal school was completed in 1962.

The presence of the whalers brought major changes to the Inuit who were attracted to the stations and employed as boat crews or hired to process the whales. Some Inuit hunted and trapped so they could trade meat and furs with the captains. The close association with the outsiders led to disease, sickness, and death among the Inuit, whose population was reduced by as much as two-thirds.

By the time that whaling disappeared around 1910, the local Inuit had come to depend upon the guns, ammunition, pots, tools, tobacco, tea, and flour they could obtain only from the outsiders. While most still resided in the many camps that dotted the margins of Cumberland Sound they traded in the Hudson's Bay Company post established at Pangnirtung. Migration from the land into the settlement increased when the school was built and hastened when an epidemic killed a number of dog teams, then the principal means of land transportation. Most federal institutions and public housing schemes were in place by the early 1970s.

Pangnirtung has nearly 1,100 residents, all but 70 of whom are Inuit. As elsewhere in the north, the population is youthful: 44 percent are children, and 55 percent are of working age—half of whom are in the wage labor force and half choose not to be. The public sector of the local economy includes jobs in administration and services related to housing, roads, the airport, health, and education. Businesses include general retail stores, fish camp outfitters, a hotel with a restaurant, coffeeshops, and taxis. As the headquarters for Auyuittuq National Park, Pangnirtung draws 25 percent more air passengers than other Arctic settlements of similar size. On nearby Kekerten Island, designated as an Historic Park, tourists can follow interpretive trails that explain the functioning of the whaling station that operated there for sixty years. Some of the people of Pangnirtung earn a living from trapping, mainly white fox, but more people hunt seals, belugas, narwhals, and polar bears, although they are limited by quotas. The settlement is also known for its arts and crafts, including carving, printmaking, and weaving, partly orchestrated through the local cooperative. Since its inception as a test fishery project in 1987, Cumberland Sound Fisheries has been developing ice fishing for turbot. The fish are caught by long lines winched up through holes in the ice from March onward, and transported in seawater to the Pangnirtung fish plant to be filleted, packaged, and flown to Montréal. Gross sales are $600,000 annually. The product is of exceptional quality, but as yet there is no hard evidence of the turbot's renewability.

The customary social facilities in Pangnirtung include a nursing station, two schools, a community hall, a school gymnasium, a library, an ice rink, and a museum.

Pangnirtung to Pond Inlet

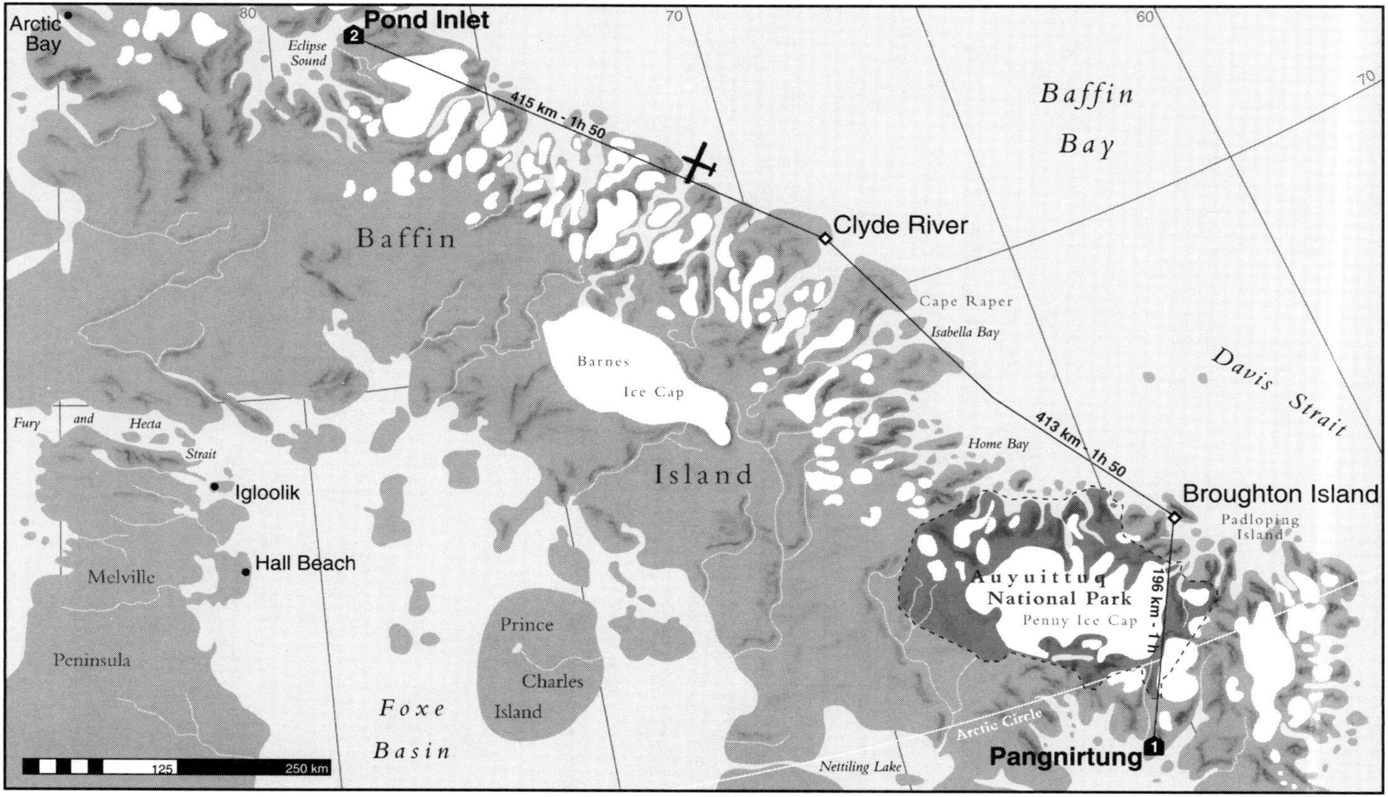

△ Day Two

PANGNIRTUNG, BROUGHTON ISLAND, ISABELLA BAY, CLYDE RIVER, AND POND INLET

Pangnirtung to Broughton Island

DISTANCE: 196 km (122 mi)
FLYING TIME: 1 hour

This route, crossing Auyuittuq National Park by way of Pangnirtung Pass, presents some of the most spectacular scenery in the north, and in all of Canada. This deep glacial trough has many tributary valleys filled with active glaciers that flow down to the main trench. Towering rock pinnacles have sheer drops of hundreds of meters, and complex geologic structures are visible on the granitic rock faces. The flight is all the more spectacular when flown within the confines of the valley, where you can look out your window and see these natural wonders pass by almost at eye level. At the top of the pass, the Penny Icecap lies like a flat, white pancake, uniting all the summits and sending its overflow tongues down between the peaks to the valleys below.

After descent from the high ground of the icecap to the north coast, the route crosses the mouths of several fiords, and the eye easily follows their lengths. Occasional icebergs stand prominently well above any sea ice that remains, or float like ships on the sea.

Broughton Island

Before the Broughton Island settlement was established in 1956–1957, the peoples of the area lived in camps up and down the coast. During the whaling period, a station was active from the late nineteenth century at a site 64 km (38 mi) north of the current town. For many years it served as a gathering place where native people came to participate in the whaling, much of which was seasonal but focused on Inuit summertime activity. After the demise of whaling, many of these people drifted north toward Pond Inlet and south toward Padloping Island.

The catalyst for the formation of the Broughton Island settlement was the construction of a Distant Early Warning Radar System station on the island in 1958. The Hudson's Bay Company established a post there two years later. In 1964 the government encouraged the people of the small settlement of Padloping Island, to the south, to move to Broughton, and these new residents provided the stable population, the current settlement.

Broughton Island is a community of about 470 people, of whom 94 percent are Inuit. As in most Inuit communities the population is very youthful: approximately 45 percent are under the age of fifteen. Several wage employment positions provide services to the settlement, but almost half of the adult population base their economic livelihood on traditional hunting and trapping. Sealskins were a major source of income until the European campaign against Atlantic sealing caused the buying public to shy away from the product; although the Inuit were not the central target of criticism, their harvest became unwanted. Recently, however, the sealing industry has revived somewhat. Local initiatives to establish a sealskin tannery that will use an environmentally safe process to manufacture sewn products are moving forward, with government support. Local women have formed the Minguq Sewing Group to make sealskin products for the tourist trade and export south.

Broughton Island is on the north side of Auyuittuq National Park and is the northern terminus of the Pangnirtung Pass Hiking

Trail. The community, therefore, is intimately involved with the policies and practices that govern the wilderness uses of the park. Near Broughton, park conditions are best suited for naturalist observations of flora and seabird rookeries. The nearby fiords and passing icebergs provide a highly photogenic landscape.

Some of the least desirable aspects of modern civilization have also surfaced at Broughton Island. Recent scientific surveys demonstrated that 63 percent of children tested had higher levels of polychlorinated biphenyls (PCBs) in their blood than the tolerable limit determined by health authorities. Moreover, 39 percent of women of childbearing age also showed levels above prescribed limits, and traces of PCBs were detected in mothers' milk.

Native peoples of the Arctic are more likely to ingest high dosages of industrial waste chemicals than non-Arctic populations for two reasons. First, the world's polar regions may be the ultimate sink for airborne PCBs, polyaromatic hydrocarbons (PAHs), and chlorofluorocarbons (CFCs). Many of the chemicals in these families evaporate at middle-latitude temperatures only to precipitate as solids in the range of 0°C to –20°C (32°F to –4°F). In other words, they are continuously mobilized in the tropics and middle latitudes and deposited at the poles where they enter the local water supply and food chain.

Second, marine mammals constitute a primary source of human food in the region. These animals concentrate the contamination at the top of the food chain. Because wild meat has long been a staple of the Inuit diet and because it is believed that communities function best when traditions are retained, these findings pose a serious dilemma. After consulting with scientists, the Broughton Island community decided that the importance of the traditional way of living and the absence of real alternatives dictated that the diet of game would continue. Further scientific research will try to increase understanding of the consequences of PCB ingestion and help determine a future strategy. It should be noted that people in other settlements who depend on wild meat and who have not been tested for PCBs are also at risk.

The biological consequences of the current human contamination levels are unclear. What is clear is that replacing a centuries-old diet of fresh food with a diet of imported, processed food would have enormously detrimental social and economic consequences, if it could be done at all. Research continues. Meanwhile, Canada has succeeded in having several of the most ominous waste chemicals found in the Arctic added to the list of substances monitored internationally under the program for Long-Range Transport of Air Pollutants, with a view to seeing them eventually reduced at the source.

Broughton Island stands on a raised beach, flanked by high, glaciated hills. The climate is similar to that of Clyde River.

Broughton Island to Clyde River

DISTANCE: 413 km (257 mi)
FLIGHT TIME: 1 hour, 50 minutes

The flight to Clyde River passes mainly off the shore of Baffin Island, and the best view is on the left side of the plane. For the first half of the journey, the route crosses Home Bay, which has several rocky, glacially scoured islands. If sea ice is present, it is interesting to observe the melting patterns or the remnant chunks that are crushed together wherever wind will drive them. At the northern end of the flight, several flat, broad peninsulas or forelands protrude from the mountain face, with fiords dividing them from one another. The flat surfaces represent preglacial and interglacial accumulations of fluvial and fluvioglacial sands, silts, and gravels that, in the last ice advance, were covered by coalescing glaciers flowing out of the mountain valleys. During the ice retreat, remnant ice blanketing the forelands protected them from shore erosion, while the active valley glaciers carved the fiords in between. Look for very large erratic rocks sitting right on the surface. It is also possible to see tundra polygons from time to time, but the rest of the surfaces, which are unconsolidated materials, have other smaller sized circles and stripes.

Isabella Bay

Nineteenth-century commercial whaling destroyed the delicate balance that existed between bowhead whales and kayak-borne human hunters. In just one century the population of these slow, copepod-eating, baleen whales was reduced from some 11,000 to the current level of 200 or so. Spending their whole lives in the Arctic—from Foxe Basin to Davis Strait to Lancaster Sound and adjacent waters—these 20-meter (66-ft), 70-ton animals developed a coat of blubber so thick that they floated when killed.

Though no bowheads have been taken commercially for more than eighty years now, the species is not recovering. Predatory killer whales, slow reproduction, and probably other factors not yet understood have kept this relic population to about 200. Wintering near the floe edge of the ice between Greenland and Baffin Island, pregnant bowheads and cows with calves follow the Greenland Current northward in late spring. They then proceed west up Lancaster Sound as far as Prince Regent Sound, using the ice as protection from killer whales. From mid-August to mid-October, mature males and calfless females congregate in Isabella Bay for the social season. They socialize, belly rub on the bottom, and breed in the shallow water off Cape Raper; they feed in the vertical gyres of the Baffin Current in nearby offshore troughs.

Bowhead whales are exceptionally sensitive to noise. Motorized boats and low-flying aircraft should stay away from Isabella Bay during the critical social period. The Clyde River Inuit were the first to recognize the need to conserve this seriously endangered stock and have recently been active in securing private and governmental support for the creation of a bowhead sanctuary in Isabella Bay.

Clyde River

When the whalers first crossed to the west side of Baffin Bay and drifted along the east Baffin shore, their first encounter with Inuit

Bowhead Whale Concentrations

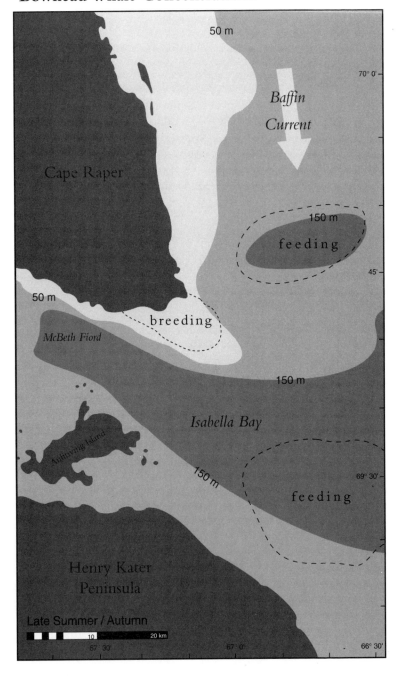

may have occurred in 1820 near Clyde River. Subsequently, most of the Inuit in the area moved to the shore to be near whaling stations, and when the whalers did not return after 1915, the Inuit suffered real distress. In 1922 the Hudson's Bay Company opened a trading post at Clyde River to exploit the fur resources there. The company relocated several families from the whale camps to the new post, which attracted former inhabitants of the region to return. Traditional hunting activity dominated the economy, with furs and sealskins being traded for the necessary manufactured goods. In 1953 the government built a Loran station at Cape Christian, 16 km (10 mi) northeast of the settlement, and an RCMP post was placed there as well. In the late 1960s, the settlement was relocated across the river mouth at the head of Patricia Bay on the east side. The Loran station closed in 1975.

Clyde River and Patricia Bay are at the seaward end of Clyde Inlet on the north side. The inlet is a typical east Baffin fiord with walls steeply descending from peaks and icecaps 1,600 m (5,200 ft) above the sea and extending inland more than 100 km (60 mi). The townsite is on a gravel beach ridge, along a level-to-gently-sloping foreshore. Surrounding it and seaward of the town is a coastal platform; in the other direction, glacially smoothed knobs and small mountains with heights of 500 to 800 m (1,600 to 2,600 ft) rise to join the mountain spine of the east Baffin coast.

The Clyde River settlement has a mean July high temperature of 7.8°C (46°F) and a mean low of 0.4°C (33°F). February, the coldest month, has a mean high of –23.8°C (–10.8°F) and a low of –27.7°C (–17.9°F). The lowest temperature recorded was –46.8°C (–52.2°F), and the warmest reached, 22.2°C (72°F). Rainfall averages 45.6 mm (1.8 in.) and snowfall 168.9 cm (66.5 in.) for a total annual precipitation of 206.4 mm (8.1 in.).

Current estimates put the population at 495, all of whom are Inuit except for ten or twelve persons. Forty-seven percent are under fifteen years of age, and half of the adults regard themselves as part of the labor force because they have worked or are looking for work. Unemployment is high—recently recorded at 40 percent. Wage jobs exist for janitors, maintenance workers, clerks, truck drivers, a mechanic or two, teachers; a few other jobs relate to providing governance and municipal services to the community. It is not enough to provide work for all who would take it. Some might go to Nanisivik on rotation. The inhabitants of Clyde River are very traditional, and the hunting of sea mammals and caribou is important both for the food it provides and the skins it brings in to be sold for cash. Furs and polar bear skins also provide some income, as do carvings and crafts. Clyde River had for some time a successful printmaking shop, but its circumstances now fluctuate. The settlement has the usual health center, a school with grades 1 through 10, and an RCMP post. The school gym serves as a community hall for meetings and movies. Sealift brings supplies from Montreal once a year in September.

Clyde River to Pond Inlet

DISTANCE: 415 km (258 mi)
FLIGHT TIME: 1 hour, 50 minutes

The route takes the traveler right along the coast, sometimes cutting directly over the peninsulas that divide the several major fiords in this part of Baffin Island. The scenery is similar to that experienced just south of Clyde River. The high mountain rim of Baffin Island is visible to the west, while below broken sea ice floats on the deep bays and icebergs emerge from a deep blue sea. There is always the chance of seeing bowhead and smaller whales from the air, although the standard flight altitude of 1,500 m (5,000 ft) usually allows only observation of the general landscape. The hills and island are rough and rocky, typical of the region, and when passing over land it is sometimes possible to see large patterns on the ground. Just before reaching Pond Inlet, the route crosses an icecap, and on the descent to the settlement the attractive mountains and glaciers of Bylot Island, to the north, offer an impressive sight.

Pond Inlet

Eclipse Sound, separating Bylot Island from the much larger Baffin Island, is the ancestral homeland of the North Baffin Inuit. The present settlement is surrounded by archaeological evidence testifying to Dorset and pre-Dorset occupancy back to the earliest migrations. The North Baffin Inuit of today are, in origin and language, related most closely to the people of Igloolik and Hall Beach, although culturally there are some distinctions.

Pond Inlet was named by John Ross when he landed on Bylot Island during his journey north to Lancaster Sound in 1818. John Pond, for whom the inlet was named, was Britain's Astronomer Royal. W. E. Parry is reputed to have visited Pond Inlet in 1820 when he was returning from his first expedition, on which he almost completed the Northwest Passage.

Soon after, whalers from Scotland and the United States began to frequent the nearby waters and they quickly drew the local Inuit into their orbit through trade and labor exchanges. The whalers encouraged the Inuit to trap foxes and wolves and to trade polar bear skins and walrus ivory. As a consequence, even people living some distance from whaling stations or ship rendezvous points learned to incorporate long trading journeys into their traditional travel cycle. The Pond Inlet Inuit learned from the whalers that other Inuit lived far to the north, across Baffin Bay in Greenland. In response, a local leader took some of his people on an epic journey in the 1870s to Ellesmere Island and across to Greenland, where they settled with the Polar Eskimo, although after a time the leader chose to return. Even today the Inuit of both places regard themselves as kin and regularly exchange visits to commemorate the nineteenth-century migration.

After the whaling era collapsed in the early 1900s, Inuit dependence upon outside goods was met by a number of independent fur traders. After the Hudson's Bay Company established a post at Pond Inlet in 1921, the independents disappeared. The following year the RCMP built a depot, and the Anglican and Roman Catholic churches established missions in 1929 and 1930, respectively.

Few sites in the Arctic are more physically attractive than Pond Inlet. The houses and other buildings climb in rows up the shoulder of a broad, rising hill to the airport, 60 m (200 ft) above the sea. The views across Eclipse Sound and northeastward across Pond Inlet to Bylot Island are splendid. Behind the settlement to the east are equally impressive glacier-bedecked mountains. The surrounding land is sculpted in typical alpine glaciation style with trough valleys and fiords. The Pond Inlet area is part of the highland rim of the Canadian Shield of east Baffin, Devon, and Ellesmere islands. The profile of Bylot Island sports peaks 1,200 to 1,500 m (4,000 to 5,000 ft) high, and the island icecap dome rises at least 1,850 m (6,000 ft) above sea level. The mountains behind Pond Inlet also carry an icecap that is over 1,500 m (4,000 ft) above the sea. Bylot Island has been designated a Migratory Bird Sanctuary to protect the snow goose colony on its southwestern lowlands. The Pond Inlet region is also noted for the vast numbers of seabirds that breed there and for the richness of marine and avian life, especially around Bylot Island and across Lancaster Sound to Devon Island—a region recognized internationally as an area of unique environmental significance.

Local records show the July mean high temperature to be 7.9°C (46.2°F) and the mean low, 1.2°C (34.2°F). In February, the coldest month, the mean high is –28.0°C (–18.4°F) and the mean low, –37.1°C (–34.8°F). The warmest temperature recorded was 20.0°C (68°F), and the extreme low stands at –53.9°C (–65°F). Precipitation is quite low for the east coast, averaging a total of 170.4 mm (6.7 in.) of which two-thirds is snow.

Pond Inlet is a large settlement with a population of nearly 900. The 50 or so whites are mostly adults employed by the government or private business. Of the 850 Inuit, 46 percent are under the age of fifteen and just over half of the working-age population is in the labor force. The unemployment rate is about 25 percent. Many Inuit choose to employ themselves in traditional pursuits or in crafts and carving, or by taking wage jobs as opportunities occur. Local hunters and trappers harvest a wide range of marine mammals, including narwhals and polar bears, within quotas. The hunting of caribou is also controlled. The hamlet has the usual array of

A prominent feature of every Arctic settlement is the satellite dish, aimed almost horizontally at a satellite 22,300 miles above the equator, which brings to the Inuit the finest cultural achievements of a world with trees, fast cars, faster men and women, and "Miami Vice." At least in Miami they don't have to improve reception by removing snow. Photograph by DIAND.

service and business establishments: general retail stores, food outlets, a hotel and a restaurant, recreation facilities, and a taxi service. The local cooperative offers many services, including the store, hotel, and fishing camp. Tourism is a growing part of the local economy, taking advantage of the area's scenic beauty and interesting wildlife. The two schools take students through grade 11, and an adult education center offers vocational and continuing education courses as well as extension programs from Arctic College.

Pond Inlet to Resolute

△ Day Three

POND INLET, KOLUKTOO BAY, NANISIVIK, ARCTIC BAY, PRINCE LEOPOLD ISLAND, AND RESOLUTE

Pond Inlet to Koluktoo Bay

DISTANCE: 128 km (80 mi)
FLIGHT TIME: 40 minutes

The route first crosses Eclipse Sound, where, if it is free of ice, you are likely to see pods of narwhals and belugas. Narwhals, up to four or five meters long, are patchy light and dark gray on top, and the males have an ivory tusk (projecting in front) of a meter or more in length. Belugas are white, although the young are gray, and full-grown they are roughly the same size as narwhals. Both are hunted by the Inuit. Their skin (*muktuq*) is much prized as food, and their tusks are valuable because they are carved or sold for ivory as souvenirs.

Flown direct, this short flight passes over recently sculpted glacial landscapes. Immediately east of Pond Inlet, where the Salmon River debouches into the sea, there is usually a summer hunting camp. South of Eclipse Sound, low Mesozoic hills north of Oliver Sound give way to mountains and fiords carved first into sedimentary rock and then, even more spectacularly, into a small granitic batholith.

If weather or the pilot's whim takes the plane north of Pond Inlet, over the Mesozoic southern plain of Bylot Island, mid-summer will reveal tightly knit flocks of snow geese waddling rapidly away from the noise of the plane. A responsible pilot will eschew too-low altitudes here. This dwarf shrub- and sedge-covered, pond-dotted plain is also a breeding ground for old-squaws, king eiders, red-throated loons, and assorted shorebirds and a feeding ground for barren-ground caribou and Arctic foxes.

Koluktoo Bay

Around the cape at the northeast end of the bay, in mid-summer, a herd of narwhals often gather to socialize. In early August of 1991 it was estimated that 200 individuals were congregated.

At the head of the bay is a beach with a barely discernable short-takeoff-and-landing (STOL) airstrip, its high, gravelly bench excitingly close to the beach's eastern end. Char gather offshore here from mid-August until the bay freezes, in preparation for their spawning run up the Robertson River.

Koluktoo Bay to Nanisivik

DISTANCE: 148 km (92 mi)
FLIGHT TIME: 45 to 60 minutes

The direct route, up the Robertson River and over the Borden Peninsula to Nanisivik, reveals a gently rising landscape of rounded surfaces and steep fluvial valleys in the uplands. The sedimentary rocks lying in distinctive horizontal layers are sometimes quite brightly colored, with rusts and reds of iron oxides. In places, erosion has produced tall cliffs with scree slopes leading into the valleys below. The upper surface of the peninsula, sometimes vegetated, is more rounded with relatively less relief.

If Nanisivik's jet-capable main airstrip is weathered in, aircraft use a more interesting and convenient STOL strip on the interfluve just east of town. The area round Nanisivik has a surface covered with blocky, fractured rock materials.

If time permits, flights down Tremblay Sound and along Navy Board Inlet to the mouth of the Maia River may provide views of herds of light-grayish harp seals. A 1991 reconnaissance flight team sighted about twenty herds, each with 15 to 200 animals, in the open waters of the inlet. When such large numbers of seals congregate, polar bears will predictably be found along the shoreline or, if present, on the ice. Following the Maia River west to return to Nanisivik provides the same general view of an occasionally dramatic fluvial landscape in the generally horizontal Archaean and Proterozoic sedimentaries mentioned above.

Nanisivik

The presence of ores along Strathcona Sound has been recognized since at least 1911, and prospecting began as early as 1937. After the Geological Survey of Canada mapped the area in detail in 1954, serious economic interest increased. In the 1970s, a body of an ore described as pyrite was measured at 120 m (400 ft) wide, 20 m (66 ft) deep, and over 3 km (2 mi) long. In 1974, the Mineral Resources International (MRI) corporation signed an agreement with the Canadian government that permitted MRI to construct and operate a mine. The government granted a subsidy and took an equity position in the mine. Underground mining began in 1976, after MRI had built a townsite. Nanisivik Mines Ltd. now has a mill capacity of 2,000 tons per day; 680,000 tons were mined in 1989. Mine reserves are estimated at 2.5 million tons of ore containing 10 percent zinc, 3 percent lead, and 45 g of silver per ton. In August and September the company ships concentrates to markets mainly in Europe. Nanisivik Mines employs 200 people.

The mine and settlement sit on a 400-m (1,300 ft) bench, 4 km (2 mi) south of Strathcona Sound and 27 km (17 mi) east of Arctic Bay. Hills and bluffs of sedimentary rock 100- to 200-m (330- to 660-ft) high surround the community to the west, south, and east. The ground is almost devoid of vegetation, except in parts of the townsite. The climate resembles that of Arctic Bay, although Nanisivik is more exposed and probably windier because of its higher elevation.

Refueling in the Arctic is not as simple as driving up to a gas station. At Nanisivik, for example, specially blended Turbo Arctic is brought in barrels by annual sealift from Montréal, trucked up 600 meters to the airstrip, and dug out as needed to be hand-pumped into the aircraft's wings by the co-pilot. R.H.I.P. Photograph by H. Swain.

When MRI and the federal government decided to proceed with development of the area for mining, they also decided to create a comprehensive mining community, with homes and town facilities for miners and their families, even though only eleven years were promised for the mine life. This strategy contrasts with the one adopted at Polaris, the only other High Arctic mine, which elected to maintain a "temporary" mode, bringing the work force to the mine on rotation. Three hundred twenty people live in Nanisivik, of whom 40 percent are Inuit and 60 percent are non-native. Two-thirds of the 110 children are Inuit. With the exception of some women, mostly Inuit, all adults work in the wage economy, and there is no real unemployment. Hunting and trapping for country food are important activities, but are pursued only after work and on weekends. Nanisivik is a company town that offers all the necessary services: airport, health center, school through grade 9, community club, gym, swimming pool, and library. The RCMP, stationed in Nanisivik, also patrol Arctic Bay. The annual Midnight Sun Marathon, run between Nanisivik and Arctic Bay on the July 1st weekend since 1979 (July 1st being Canada Day), is famous among runners the world over. The longest course, 100 km (62 mi) includes a grueling 500-m (1,640-ft) difference in elevation as the demented run back and forth between the two settlements.

Arctic Bay

The archaeological record shows that Paleo-Eskimo people occupied north Baffin Island from their earliest wave of settlement around 2000 B.C. The known Paleo-Eskimo sites are concentrated around Pond Inlet on the island's east side and toward Igloolik to the south, but Admiralty Inlet, near which today's settlement of Arctic Bay is situated, was certainly within the hunting zone. North Baffin was also centrally located within the core zone of the Dorset people, and remains of Thule culture, the whaling culture that spread eastward quickly across the Arctic about A.D. 1000, are also found along the inlet. In historic times, whalers hunted in Admiralty Inlet and traded with local people; indeed, the bay on the south side of a narrow isthmus was named after the whaling vessel *Arctic* in 1872. The Inuit name, Ikpiarjuk, meaning pocket or bag, describes the shape of the bay. The Hudson's Bay Company built a post at the current townsite of Arctic Bay in 1926, attracting more people to Admiralty Inlet; the post has been in operation more or less continuously since that time.

Arctic Bay is laid out with its buildings facing southeast to the bay. It is three streets deep and rises on flights of old beaches up a moderate slope behind the town. Above the settlement is the saddle of the isthmus, 150 m (500 ft) high, that drops north to Victor Bay 1.5 km (.9 mi) away. To the west is the wedge-shaped Uluksan

Peninsula with hills as high as 440 m (1,450 ft). East is King George V Mountain, the highest point at 577 m (1,893 ft). The area around Arctic Bay is noticeably barren of vegetation. The bedrock is a mixture of sedimentary and volcanic rock, and there is also a distinctive soapstone called *kooneak* that is quarried and carved by the local people. Lead/zinc deposits in the area are the basis for the mining town Nanisivik, 27 km (17 mi) away.

Arctic Bay records a mean high temperature of 9.5°C (49.1°F) in July; the mean low for the same month is 1.7°C (35.1°F). It is coldest in February, with the mean temperature ranging from a high of −27.4°C (−17.32°F) to a low of −35.1°C (−31.2°F). The extreme low recorded is −49.4°C (−56.9°F), and the warmest day on record reached 23.9°C (75°F). It is quite dry in Admiralty Inlet. The Arctic Bay weather station normally records about 51.7 mm (2 in.) of rain—rain falls an average of nineteen days a year—and 71.5 cm (28.1 in.) of snow, falling over an average of sixty days. Total precipitation is but 118.3 mm (4.7 in.). Winds at the surface reflect the orientation of Admiralty Inlet and blow prevailingly from the north.

The population of Arctic Bay is 550, of which all but about 20 people are Inuit. It is a settlement full of young people; almost half the population are younger than fifteen years. Fifty-six percent of Arctic Bay adults are in the labor force, and the employment level is fairly high. Most residents work in the settlement, but others go to Nanisivik to work on rotation at the mine. Hunting is important and seals are plentiful, but caribou, polar bear, and musk ox hunting is limited by a quota system. There is some trapping of white fox and wolf. The settlement infrastructure includes a community health center, a school through grade 10, the Cooperative and Northern stores, offices, a garage, and a sod museum.

Arctic Bay to Prince Leopold Island

DISTANCE: 212 km (132 mi)
FLIGHT TIME: 1 hour

Admiralty Inlet separates Arctic Bay from the Brodeur Peninsula. Large numbers of belugas and narwhals inhabit this broad stretch of water. The eastern edge of the Brodeur Peninsula is a cliff-like shore rising to 500 m (1,600 ft) and showing scree slopes with a narrow beach. From the cliff tops the land grades gently to the much lower western edge of the peninsula. Sedimentary limestones and shales produce a more softly rounded surface, although cliffs form along deep valleys. Tundra polygons and nesting waterfowl can be seen along any of the lakes. Prince Regent Inlet separates the Brodeur Peninsula from Prince Leopold Island.

Prince Leopold Island

This steep-sided *tarte tatin* of Silurian and Devonian limestones lies in Lancaster Sound off the northeast tip of Somerset Island. STOL aircraft may land on the gravel spit on the southeast tip, the only bit of lowland on the island, or atop the cliff above it. Either way, the landing is likely to catch the breath of the uninitiated.

On the eastern strand of the spit are the remains of some Thule houses. One in particular seems well preserved, with its two-level interior, low entrance tunnel, and lamp niche mostly intact. A whale vertebra underlines the lesson of the sere and barren shingle: These folk were not salad eaters.

From the top of the cliffs that you can best see what's special about Prince Leopold Island. If you are very careful on the frost-shattered limestone a thousand feet above the sea and are a stranger to vertigo, look down on the nests of 75,000 northern fulmars, 29,000 black-legged kittiwakes, 86,000 tick-billed murres, 2,000 black guillemot, and occasional glaucous and Thayer's gulls. Dovekies and ivory gulls feed in the neighborhood but do not

Seabird paradise. The eastern side of Prince Leopold Island falls a thousand feet into Lancaster Sound. Tens of thousands of seabirds breed in safety on these cliffs, feeding in the nutrient-rich upwelling waters of the sound. Photograph by H. Swain.

breed here. By early August, most of the birds will have left, although hundreds will still be present.

The seabirds flock to Leopold's cliffs, and to other dizzying locations on the outer coasts of Lancaster Sound, to take advantage of the upwelling of nutrient-rich abyssal waters swarming with the plankton that support fish, seals, and indeed all of the Arctic's larger fauna.

Prince Leopold Island to Resolute

DISTANCE: 171 km (106 mi)
FLIGHT TIME: 50 minutes

You will cross over Barrow Strait, part of the Northwest Passage. If ice is present, watch for seals and polar bears. Otherwise, look for whales in the open water.

Resolute

Cornwallis Island, on which Resolute is situated, was not inhabited in historic times, although Dorset and Thule archaeological sites on the island testify to a time when the climate was warmer and more open waters were frequented by the baleen whales. The coast was first known to be seen by Europeans in 1819 when W. E. Parry had an unobstructed passage through the channel that bears his name, as far as Winter Harbour on Melville Island. Resolute Bay was named after the *H.M.S. Resolute,* which wintered at nearby Griffiths Island in 1850–1851 during the search for Franklin. Crewmen on sledge journeys from search ships at Griffiths Island and Assistance Bay, a little east of Resolute Bay, examined and mapped most of the coastlines in the area in detail. *Resolute* returned to the Arctic the following year to continue the search but was unable to beat the ice when heading home and spent the winter of 1853–1854 off Bathurst Island. In the spring of 1854, the ship, still stuck in the ice, was ordered abandoned, prematurely it turned out, by the senior command. The crew walked east over the ice to be rescued and returned to England. The *Resolute* floated free the next summer and drifted out to Davis Strait where it was claimed by an American whaling ship. The U.S. government refitted the ship and presented it to Queen Victoria. When it was later taken apart, timbers from the ship were used to build a table presented as a gift to the President of the United States. The table still stands in the White House.

Resolute is really an artificial settlement. The site was selected in 1947 as the control center for five manned weather stations in the Queen Elizabeth Islands. An airstrip was built and the station opened, to be jointly operated by Canada and the United States. The Royal Canadian Air Force also established a base there in 1949. Since that time, the station has kept an unbroken record of surface and upper-atmosphere weather as well as other data. The main purpose for constructing the High Arctic weather stations was to provide forecasters with the weather observation data essential for the development of transpolar commercial flying. Later, the weather station became the sole responsibility of the Canadian government.

In 1953 an Inuit community was established in Resolute. Faced with declining living standards of Inuit in the Hudson Bay area, the government, at the suggestion of the RCMP, considered relocating Inuit groups to areas where game was more plentiful. Eventually the government decided to relocate them to High Arctic RMCP post sites. In the summer of 1953, three families (fourteen people) from Port Harrison, Québec, and one family (five people) from Pond Inlet, plus an RCMP constable, were left on the beach at Resolute Bay by the patrol vessel *D'Iberville*. In 1955 three more families (twenty-one people) came from Port Harrison. Between 1968 and 1972 a few families from Fort Ross, Somerset Island, and Great Whale River moved to Resolute. The current townsite was laid out in 1977.

Resolute Bay is three miles across at its mouth and extends two miles inland. It has a gravel shore made up of closely spaced, raised beaches that look like flights of contour lines rising inland. Vegetation is extremely sparse, and much of the surface is covered with frost-fractured limestone that resembles broken crockery. Frost boils, stone nets, and other forms of patterned ground are easily visible. Behind the settlement is Signal Hill, a prominent landmark about 200 m (650 ft) high. Much of the upland part of Cornwallis Island is at this elevation.

In July the mean high temperature is 6.8°C (44.2°F) and the mean low is 1.4°C (34.5°F). It is coldest in February, with the mean high at −29.6°C (−21.3°F) and the mean low at −36.8°C (−34.2°F). The coldest temperature ever recorded was −52.2°C (−62°F) and the highest was 18.3°C (65°F). Precipitation is very light, averaging 131.4 mm (5.2 in.) a year. Most of it comes as snow because temperatures are below freezing ten months of the year. Even so, the snow cover is thin on exposed surfaces, although it is packed hard in drifts in lee locations by the wind, which seems to blow constantly from the north and northwest. Because it is poleward of the Arctic Circle, Resolute has continuous daylight during May, June, and July and experiences the polar night of 24-hour darkness in November, December, and January.

Cornwallis Island was occupied by Thule Eskimos until about 500 years ago. Refurbished Thule houses are within walking distance of the current settlement. Today, there are 120 resident Inuit and a non-Inuit population of 40 to 50, mainly working adults. From the hamlet's beginning, Resolute's two communities—the Inuit settlement and the base at the airstrip five miles away—have maintained a certain cultural and social distance. Of the 70 or so adult Inuit, about half are in the labor force; some are employed at the base, but the majority work in the settlement. Wage employment is available in municipal and other government institutions such as the school and the nursing station, as well as at the Cooperative and in private businesses. Hunting and trapping are still important to the local Inuit. They can freely hunt white foxes, seals, and can take belugas, musk oxen, and polar bears under quota controls. The usual settlement services are available from the Co-op store, the hotel has a restaurant, and recreation activities are available at the Community Hall. Resolute has also opened an aquarium that displays Lancaster Sound marine life. A surprising amount of tourism is focused here, especially through outfitters that promote wilderness travel.

For the outside world, Resolute acts as the hub of the High Arctic. It serves as the base for scientific activity through the Polar Continental Shelf Project coordination center. Airlines maintain depots in Resolute, and both scheduled and charter flights service the area north of 75° latitude for passengers engaged in resource exploration, scientific fieldwork, and government activity. Obviously, the traffic generated by field parties is seasonal.

Resolute to Bent Horn, to Resolute

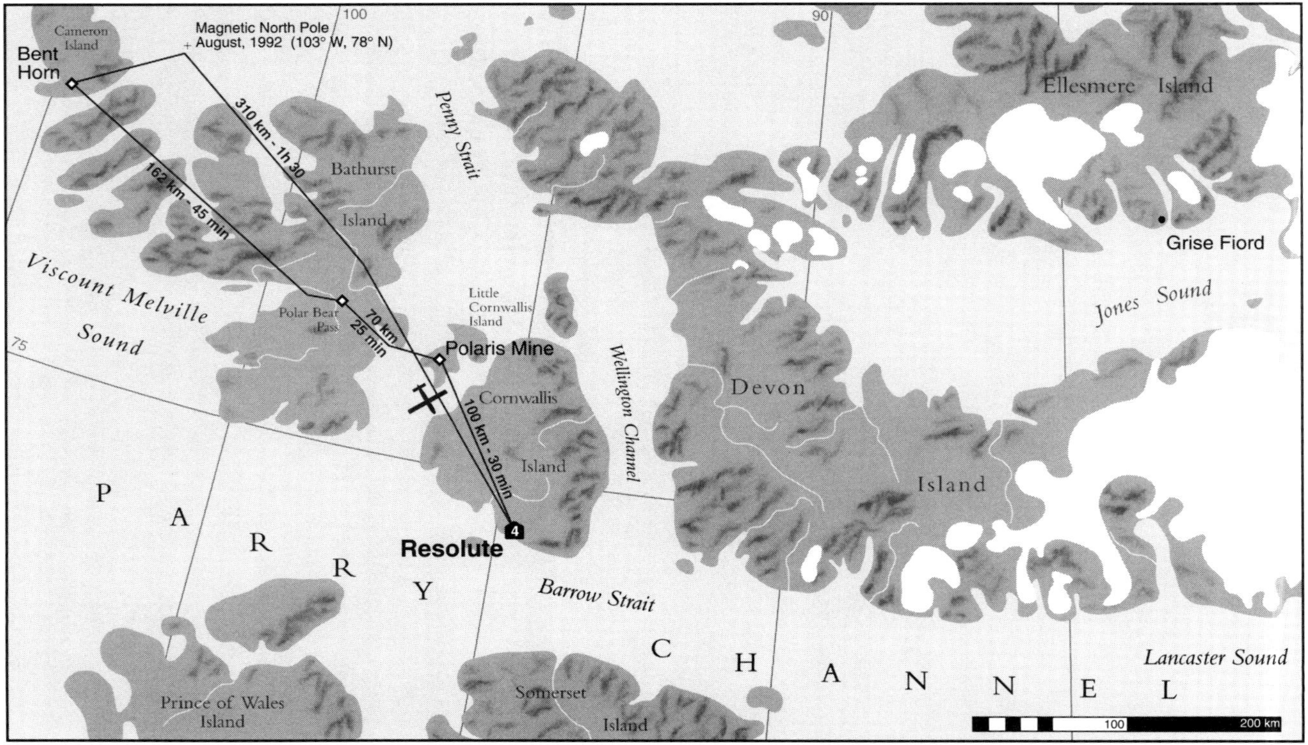

⟁ Day Four

RESOLUTE, POLARIS MINE, POLAR BEAR PASS, BENT HORN, AND MAGNETIC NORTH POLE

Resolute to Polaris Mine

DISTANCE: 100 km (62 mi)
FLIGHT TIME: 30 minutes

This short journey runs northwest across the western side of Cornwallis Island to Little Cornwallis Island. The geologic base of the larger island is sedimentary rock, primarily limestone. Occasional cliffs of capped bedrock appear, particularly along the deepest river valleys, but otherwise the surface is moraine and deeply weathered bedrock. Flanking all the shorelines are flights of raised beaches that are continuous except where cut by stream erosion. Almost everywhere on the uplands light and dark solifluction patches streak the landscape like paint poured over it, flowing as water would down over the surface. One can see patterned ground everywhere, sometimes large tundra polygon cracks, and often circles, polygons, and downhill stripes. Vegetation is very sparse, except in a few broad valleys and other low, protected areas where snow accumulation provides enough moisture to nourish it. The barrenness indicates that large mammals are not plentiful, although both caribou and musk ox are occasionally sighted.

Polaris Mine

The Polaris Mine on Little Cornwallis Island is the northernmost metal mine in the world. At latitude 73°23′N, it produces zinc/lead concentrates at a rate of 270,000 tons per year. The operation is thoroughly modern and is designed to extract ore from an ore body encased in permafrost. Special operational problems result from the Arctic conditions of low temperature and continuous winter darkness. Indeed, to maintain the ore's competence during the short summer, special efforts are made to keep the ore body frozen. Ventilation air is chilled at the mine portal, but even so, moisture precipitates and refreezes as ice crystals on the walls and ceilings of stopes, adits, and rooms, making the whole operation look more like a salt mine than a lead-zinc body. Gangue is pumped as water-based slurry to a disposal site in a nearby meromictic (layered-brine) lake, and must be specially heated and insulated along the way. Permafrost enhances the competence of mine backfill, however. The sand slurry used as backfill freezes in place, expanding slightly as it does so. The mined concentrates, stored in a big shed on the site, are shipped during the brief open-water season of August and early September, mostly to Antwerp in Belgium. All bulk materials and heavy equipment are also delivered to the mine at that time.

The ore body at Polaris is a rich one, with 14 percent zinc and 4 percent lead. After ten years of production, 12 million tons of the body remain, an amount that can last another twelve years. The mill can process 2,800 tons per day and operates continuously all year around.

The first signs of mineralization at the site were noted in 1960, but a decade passed before geologic study made clear the serious economic potential of the area. Another decade passed before operations began, during which time some very innovative concepts emerged for the configuration of the mine facilities. The most imaginative idea accepted was to construct the mill as a self-contained plant, with maintenance shops, the power plant, and fuel storage for the whole operation built under one roof—or, more accurately, built into a single floating barge. After construction in Québec, the barge, as long as a football field and six stories high, was towed from Trois Rivières for eighteen days to the prepared site. The lagoon dock was then sealed, its water pumped out, and the barge-mill made ready to operate by August of 1981. The first ore went through the concentrator in November of that year, beginning production.

A large accommodation block at Polaris Mine contains 200 rooms and sixteen suites; eating and recreation spaces, including a swimming pool; and all offices. The entire operation is self-contained; that is, the mine management assumes full responsibility for all aspects of the enterprise. Polaris has 200 employees, 20 percent of whom are native northerners. Because there was never any intention to establish the mine as a community, all employees live and work at Polaris on a rotating basis. Native employees work for six weeks and go home for four weeks. Others hold to a nine-weeks-in, three-weeks-out rotation scheme.

Polaris Mine to Polar Bear Pass

DISTANCE: 70 km (43 mi)
FLIGHT TIME: 25 minutes

Polar Bear Pass, on Bathurst Island, is west of Polaris across Crozier Strait. The trip is an opportunity to see marine mammals and sea ice. There is a haul-out for male walrus on the Bathurst shore opposite Kalivik Island, and if ice is present, seals often bask upon it. When these animals gather, there is always a chance to spot a polar bear.

The shores of Bathurst Island are very similar to those of Cornwallis Island since their geologies and physiographic histories are similar. In bright sunlight, the land is colorful—sand tans, browns, and blacks, with patches of rusty-red looking like dye running down the hills. The lowlands display patches of vegetation.

Polar Bear Pass

This low-lying pass on southern Bathurst Island is one of those rare concentrations of ecological richness that occur when soils and microclimates conspire to favor, if only slightly, one site among hundreds. Established as an ecological reserve in 1982 as part of the International Biological Program, Polar Bear Pass has had only limited predation in recent times. But during the decade following the movement of Inuit to Resolute Bay in 1953, hunting pressure was relatively high. Thereafter, the growing wage economy in Resolute, plus Inuit agreement in 1973 to cease hunting the rare Peary caribou altogether, led to an easing of human predation. A seasonal research station was established at Polar Bear Pass in 1968.

Geologically, Polar Bear Pass is a thrust fault in Devonian limestones and dolomite, which are exposed as stacks on the northern margin. Although the pass is called an Arctic oasis, such a statement is relative: Winter temperatures average −35°C (−31°F), while temperatures in the three warmest months average −1°C to 4°C (30°F to 39°F). The persistent moisture in the poorly drained lowlands, together with locally high insolation, compensate for the otherwise desert-like level of precipitation. The disorganized drainage is, as in so many places in the Arctic, a consequence of the recent emergence of the landscape from below sea level coupled with underlying permafrost and exceptionally low fluvial energy levels. Seashells recovered from a central area of the pass have been dated at 6500 B.C., a measure of the speed of isostatic rebound. Around the ponds and lakes, rich sedges and grasses contribute annually to a thickening layer of peat. These saturated meadows develop tundra polygons and other forms of patterned ground.

On the terraces and raised beaches is an almost continuous mat of lichens, interspersed with mosses, sedges, grasses, various flowering plants—especially saxifrages and mountain avens—and Arctic willow. Vegetation becomes sparser with altitude, and the patches of avens, saxifrages, and Arctic poppies grow ever more scattered.

Long-legged Arctic hares do not change color during the brief summer, though the first coat of spring-born young is a grayish color less likely to attract the attention of Arctic foxes and wolves. Photograph by DIAND.

This rich natural garden attracts birds—fifty-four species have been recorded, of which thirty breed here—herbivores, and predators. Snowy owls appear in late winter to prey upon lemmings—and the lemmings, who scurry through their well-lit, insulated snowdrift colonies, provide the raw material for a characteristic orange lichen found atop isolated boulders where the birds transform lemmings into owl pellets. A few other species, notably snow buntings, show up by April, followed by dozens more when the melt gets underway in June. Rock ptarmigan live here year-round. Dust from the wind-swept hilltops settles on the valley snows, lowering the albedo, which in turn allows warming that activates wintering invertebrates, a source of food for red knots, black-bellied plovers, sanderlings, purple sandpipers, and other shorebirds.

The availability of food in this area, in contrast to the prolonged winter conditions of the bare surroundings, also attracts several hundred brants, king eiders, snow geese, and jaegers. The food supply is vital to early-arriving species that must replace lost energy to fly farther north and to form eggs.

Eleven mammal species, six of them marine, have been recorded here. Lemmings and Arctic hares are prey to Arctic fox, and wolves pull down weakened musk ox and Peary caribou. Polar bears hunt mainly ringed and bearded seals but are otherwise opportunistic. Polar bears can be seen most times of the year, crossing from one inlet to the other, but are especially common in tourist season, August and September. On the eastern shore, opposite Kalivik Island, is a well-known walrus haul-out. The nearby waters are rich in narwhal and beluga and harbor an occasional bowhead whale.

Recent human predation began only in 1953, and habitation is mostly confined to the visits of scientific parties. There is much evidence of Dorset and Thule occupation, at least up to the Little Ice Age of the fourteenth century. According to the Wildlife Service's management plan, the most prominent sites are of Thule origin and contain the remains of semisubterranean winter dwellings. These houses, constructed of sod and stone, use whale ribs and mandibles as roof supports. Summer tent rings, meat caches, burial cairns, stone traps, and kayak and *umiak* supports are often associated with the sites. Arctic small tool tradition sites are also fairly common; they are characterized by patches of lusher vegetation, circles of rock, and pieces of weathered bone.

Polar Bear Pass to Bent Horn

DISTANCE: 162 km (101 mi)
FLIGHT TIME: 45 minutes

The flight from the pass crosses northwest over Bathurst Island to Cameron Island where the oil well at Bent Horn is located. The route crosses the southern part of a fold belt, and sedimentary structures show parallel lines of different rocks lying across the general pattern. Bays and island shapes pick up the orientation. The ground surfaces are unconsolidated tills and weathered bedrocks, and there is a good deal of patterned ground.

The North Magnetic Pole lies in this general area.

Magnetic North Pole

Unlike Earth's rotational pole at 90° N, the magnetic pole is an unreliable sort: It changes position in response to magnetic distortions probably caused by convection in the Earth's core. The pole was first located and measured in 1831 when James Clark Ross found his compass dipping vertically downward on King William Island. It has lately been wandering through the Queen Elizabeth Island and in mid-1992 will be around King Christian Island at approximately 78° north latitude, 103° west longitude. Travelers who want to be able to claim they have been to the North Pole can save the miseries and expense of an over-ice trek from Ward Hunt Island north of Ellesmere Island by choosing a comfortable beach on one of the Queen Elizabeth Islands and waiting the prescribed period of time. Two Canadian deputy ministers decided that Bent Horn was close enough for government work.

Bent Horn

Arctic exploration has always been driven by people with dreams, and the search for hydrocarbons in the Mesozoic sediments of the Sverdrup Basin islands was no exception. Charles Hetherington, the driving force behind Panarctic Oils Ltd., a risk-sharing vehicle for a number of oil companies, has put geological imagination, enthusiastic salesmanship, and the Canadian tax structure to work over the last quarter century. At first it was gas: on Melville Island in 1968, King Christian Island in 1970, Ellef Ringnes Island in 1971, and the Sabine Peninsula of Melville (again) in 1972. These and later large gas discoveries stimulated the dream—not to be realized for a generation yet, though that was hardly the view in the heady years after the OPEC oil crisis—of shipping liquefied natural gas to Europe in giant, ice-strengthened tankers, escorted by monster icebreakers.

Hetherington first discovered oil on the southwest coast of Cameron Island at a site now called Bent Horn A-02. Bent Horn oil is remarkably light—so light that it can be burned directly in the 2,300-hp MAK diesels at Polaris Mine having received no treatment other than simple dewaxing. Following experimentation, however, use of the oil was discontinued because its high volatility constituted an explosion hazard.

Bent Horn oil was simply a curiosity until 1985, when Panarctic installed production and storage facilities at A-02. In August of that year, 100,000 barrels were loaded into the *MV Arctic* for the Petro Canada Montréal refinery. One or two shiploads a year have been transported annually since then, with unrefined trial lots going to the Polaris and Nanisivik mines. A couple of Arctic communities have tried to use the oil in their community generators.

Bent Horn is worth visiting to learn how easy it can be to exploit oil in the Arctic. Despite the expensive fly-in camp, the long period between investment and cash flow, the brief shipping season, and the concomitant need for on-site storage, Bent Horn A-02 produces a valuable and tractable oil from a dry-land site, requiring no adventurous offshore technology, and within a few hundred yards of deep water. At a selling price of $20 a barrel, however, the 2,257 billion cubic meters of gas and 686 million cubic meters of oil that the Geological Survey of Canada estimates is yet to be discovered in the Sverdrup Basin is likely to remain there for a long time yet.

△ Day Five

RESOLUTE, BEECHEY ISLAND, TRUELOVE LOWLANDS, AND GRISE FIORD

Resolute to Beechey Island

DISTANCE: 113 km (70 mi)
FLIGHT TIME: 35 minutes

This short journey along Barrow Strait and the south side of Cornwallis Island crosses Wellington Channel to the southwest tip of Devon Island. The Cornwallis shore is a series of raised beaches rising to an upland, but without really dramatic topography. Halfway along the coast is Assistance Bay, the wintering harbor for Captain Penny on the Franklin search of 1851–1852. The east side of the island is a high—200 m (650 ft)—limestone cliff from which a major landslide left a sheer drop. The landslide had not occurred in 1819 when Parry passed west, but was observed in the 1850s when the Franklin search expeditions came through. Across Wellington Channel, the coast of Devon Island shows some cliffs.

Beechey Island is really a tombolo. Storm beaches, both ancient and modern, have built up as sea levels have fallen, forming a solid, connecting gravel bar. It is on this beach material that the graves and monuments of the Franklin expedition are found.

Resolute to Grise Fiord, to Resolute

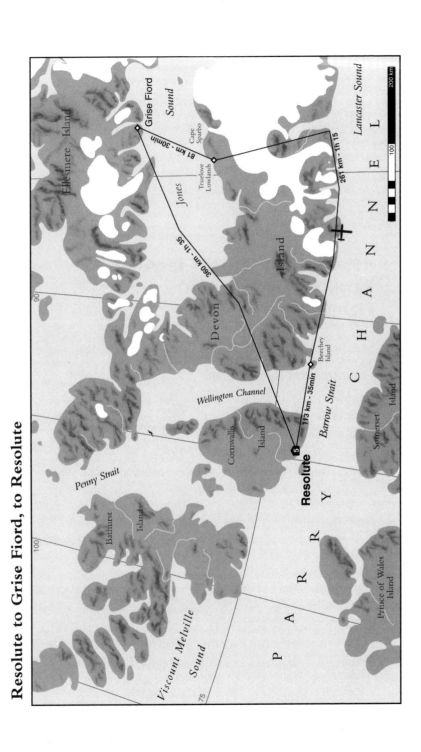

Beechey Island—Necropolis of the North

The summer of 1845 was harsh and brief, even by the standards of Lancaster Sound. Sir John Franklin's third and last expedition in search of the Northwest Passage was halted by ice at the western end of Devon Island. With his ships *Erebus* and *Terror* and 129 men, he was beset in September in the bay just east of Beechey Island. On 1 January 1846, the first crewman, John Torringon, died. He was followed on January 4th by John Hartnell and on April 3rd by William Braine. Five crewmen in all are buried in the permafrost of the shingle beach.

The subsequent history of the miserable decline of Franklin's failing band has been told many times. Recent research has added to the traditional tale of scurvy and starvation a new horror: lead poisoning from the tinned foods supplied for the voyage at, of course, the lowest bid. Torrington, already weakened by a history of lung disease, was killed by pneumonia—but not before ingesting so much lead that his hair registered an incredible 600 ppm. Mass spectometry confirmed that the source was the 90:10 solder that sealed the *Terror's* food tins.

Some yards vaguely to the northwest of the Franklin graves is the stretch of barren beach the desperate men called "the kitchen garden." A mile south are relics of one of the many expeditions sent in search of Franklin. In 1854, Captain Francis Leopold M'Clintock built the substantial, if drafty, storehouse, the ruins of which, along with rusted bully beef tins, barrel staves, and hoops, now litter the raised beach. Above is the monument to the departed hero sent by his long-suffering widow with Captain Hartstene on the U.S. yacht *Fox*. Hartstene got only as far as Disco Bay, Greenland, where M'Clintock later picked up the stone and installed it in its current location.

Now a lonely monument to the lost sailors is one thing, but others have been moved to add their own calling cards. A nearby wooden cairn contained, at last count, five bronze tubes containing records of ships' visits, a gallon jar full of notes, and two ships' plaques. Monuments have been built by the Franklin Probe

Society, by the families of two northerners who chose to be buried there in the 1970s, and even by royalty. The Prince of Wales memorialized his own visit in 1975, as did Princess Margriet of the Netherlands "and her husband, Mr. Pieter Van Vollenhoven" in 1978. The Dutch, like the English, apparently need to learn something about buying from the lowest bidder, as the concrete of their monuments is fast returning to gravel.

Should these relics and this ghostly place be protected from tourists and other vandals? Since 1988, the spar that served as McClintock's flagpole has fallen, and souvenirs have been carried away by the thoughtless.

Beechey Island to Truelove Lowlands

DISTANCE: 261 km (162 mi)
FLIGHT TIME: 1 hour, 15 minutes

The flight may either cross Devon Island or wind around the coast, depending on weather. The coast is composed, in part, of high cliffs up to 200 m (650 ft). The surface of the island interior has a general elevation of 500 m (1,600 ft). The western half of the island is sedimentary with a fairly horizontal attitude. The land is trenched by several rivers, exposing cliffed margins and braided streams. Its surfaces show plenty of evidence of downslope movements of weathered debris in the form of solifluction. Covering the eastern quarter of the island is its icecap which has a summit elevation of almost 2,000 m (6,500 ft). In summer you can sometimes see meltwater lakes on the glacier which have a remarkable turquoise-blue color.

Truelove Lowlands

On the north side of Devon Island are five large, lake-strewn, vegetationally rich meadowlands. Truelove is the farthest east. In 1960 the site was selected as a research station because it offered great opportunities for interdisciplinary studies. Scientists were attracted to the great variety of landscapes that are fairly close at hand and that contain glacial evidence and biological activity in both terrestrial and marine environments. Gradually, over the years, researchers have built up an impressive database on many interrelated subjects. Together, the data will contribute to a comprehensive understanding of the ecotone. Recently, for example, field parties have investigated the nitrogen cycle in an environment that precludes nitrogen-fixing plants; geomorphological work has been done on lake formation; other geologists have assessed shore processes; and zoologists have conducted studies on entomology and large mammal biology. Early Paleo-Eskimo sites are available for study as is evidence of later Inuit history. Truelove is also an excellent site for instrumented winter observations because it is isolated from settlements. Other projects have related to oceanography in Jones Sound and to climatic change.

The camp at Truelove is operated by the Calgary-based Arctic Institute of North America and it can accommodate up to thirty researchers. It operates from the end of May until August.

Truelove Lowlands to Grise Fiord

DISTANCE: 81 km (50 mi)
FLIGHT TIME: 30 minutes

This part of the journey crosses Jones Sound, which usually clears of ice in summer and in some locations has polynyas that concentrate marine life in early spring. Narwhals and belugas are often present.

Grise Fiord

At latitude 76°25′N, Grise Fiord is the northernmost settlement in Canada. People of the Pre-Dorset, Dorset, and Thule cultures successively occupied sites in the area, but 500 years ago cooling temperatures apparently caused the Thule people to withdraw south of the Parry Channel. The site was not reoccupied until modern times.

The Norwegian explorer Otto Sverdrup mapped the area during his 1898–1902 expedition and named Grise Fiord (*grise* means pig in Norwegian). In 1953, the Canadian government sponsored relocation of four Inuit families comprising twenty people from Port Harrison, Québec, and one Inuit family comprising five persons from Pond Inlet to the new camp.

Relocation of Inuit to the High Arctic had been tried before. Out of concern for Canadian sovereignty in the 1920s, in light of an unresolved Norwegian interest in the "Sverdrup Islands" and of American scientific and exploratory activities north of Baffin Island, the government placed RCMP posts in the north. Posts were built at Craig Harbour on southeast Ellesmere Island in 1922 and at Dundas Harbour on southeast Devon Island in 1924. In 1926, the Craig Harbour post was relocated to the Bache Peninsula on the east side of Ellesmere Island, at latitude 78° N; in 1933 the Bache post was closed and the Craig post reopened for a while. For each post, the police recruited an Inuit family from Etah, Greenland, to help with chores and dogsled patrols. Ottawa's sovereignty concerns subsided in 1930 with the purchase of Norway's interest.

Meanwhile, the Hudson's Bay Company agreed to build a trading post at Dundas Harbour at its expense and took responsibility for moving Inuit trappers from south Baffin Island where hunting was causing pressure on game and the Inuit suffered general economic distress. The company moved twenty-two Inuit from Cape Dorset, twelve from Pangnirtung, and eighteen from Pond Inlet, along with their 100 dogs, many kayaks, and other wordly possessions, to Dundas Harbour in 1934. Ice conditions and winter shore travel made life so difficult for the settlers, however, that the company closed the post in 1936 and moved the group to Arctic Bay. The next year the company relocated the same group to Fort Ross at the south end of Somerset Island. Some Cape Dorset relatives came to join the group at Fort Ross, but unreliable sea access made it hard to resupply the place, and in 1947 the post was closed and the Inuit moved again, this time to Spence Bay.

The relocation of Inuit families to Grise Fiord and Resolute at the same time was again an RCMP suggestion. On this occasion, however, the government not only supported relocation, but also paid for it. One purpose of the move was to provide new and better hunting and trapping grounds for people living in what was described as overpopulated and game-depleted districts east of Hudson Bay. Another rationale held that the officers of the RCMP post who had relocated to Craig Harbour again in 1951 would benefit by having native people nearby. Once the Inuit landed at Craig, the police determined that the migrants should set up their camps on Lindstrom Peninsula at Grise Fiord, forty miles (64 km) distant. In 1956, the police closed Craig Harbour and removed themselves and their small store to a site on the east side of Grise Fiord opposite Lindstrom Peninsula. A few more families came from Port Harrison, Pond Inlet, and Pangnirtung to join what were essentially two separate camps, one of Hudson Bay Inuit and the other of Baffin Island people. The federal school was built next to the RCMP post. With the establishment of federal housing programs at Grise Fiord, the settlement was consolidated at its current location around the school after 1964.

In recent years, issues surrounding the original colonization have led local people to express their discontent. They maintain that they did not participate fully in the decisions of the time and that, when they arrived at the new locations, they were badly equipped to face an entirely new environment with undiscovered hunting areas and a winter with polar night. A number of people complained about the government's failure to honor its promise to repatriate any families that elected to return home after two years. Some Inuit today also allege that the colonization efforts were aimed primarily at securing Canada's claim to the Arctic Islands through placing native settlements on them, and that the Inuit people thus served as pawns in the international sovereignty game. The government's position is that the sovereignty issue was not involved in the relocation policy, but that grounds may exist to acknowledge some neglect to provide adequately for the migrants during the first winter and to follow through with offers to repatriate them. The Inuit are seeking redress.

A settlement's polar bear quota can provide thousands of dollars of revenue from fees charged to affluent hunters from Europe, Latin America, and the United States. This Grise Fiord polar bear's skin has recently been removed and it is being prepared to decorate the distant home of a wealthy hunter. Photograph by H. Swain.

The townsite, set on a rising slope by the sea, faces south. It is surrounded inland by glaciated knobs with small icefields 800 m (2,600 ft) high and a larger icecap at 1,300 m (4,300 ft) in elevation. Grise Ford itself is 3 km (2 mi) across and extends north, inland, for 40 km (25 mi). In July the mean high temperature is 10°C (50°F) and the low is 2.2°C (36°F). February has a mean high temperature of −27.9°C (−18.2°F) and a low of −35.3°C (−31.5°F). Consistent with very dry conditions in the High Arctic, annual rainfall averages 0.0 mm and snowfall is 15.2 cm (6 in.) for a total precipitation of 15 mm (5.9 in.).

According to the 1986 census, 105 Inuit people lived in Grise Fiord. Two years later, the NWT government estimated the population at 86 people. The decline is caused by out-migration. Some people have moved back to Inukjuak (formerly Port Harrison) and Pond Inlet to be reunited with their families. The real dilemma today, however, is faced by younger residents who were born and have lived all their lives in Grise Fiord and who cannot really think of Hudson Bay as their home. Those who have remained have employment available to them providing government and local services, but there is also a strong relationship to the land. Local hunters take seals, beluga, and walrus, and hunters provide income by trapping white foxes and hunting musk ox and polar bears under quotas. Because it is Canada's northernmost community and has such a naturally beautiful setting, Grise Fiord also has a modest but valuable tourist trade. The settlement contains a school for students up to grade 9, a nursing station, a small community hall, cooperative store and hotel, and a police post.

Grise Fiord to Resolute

DISTANCE: 360 km (224 mi)
FLIGHT TIME: 1 hour, 35 minutes

The return journey passes over much of the same landscape seen on the trip to Grise Fiord.

Resolute to Eureka

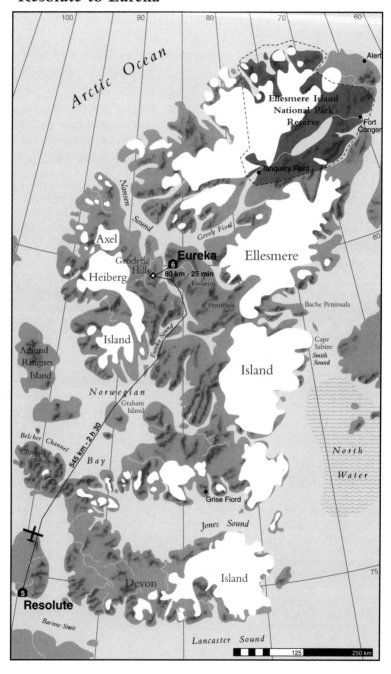

△ *Day Six*

RESOLUTE, GEODETIC HILLS, AND EUREKA

Resolute to Geodetic Hills

DISTANCE: 545 km (339 mi)
FLIGHT TIME: 2 hours, 30 minutes

This long flight north crosses the middle of Cornwallis Island, the narrow part of the Grinnell Peninsula of Devon Island, Belcher Channel, Graham Island, and Norwegian Bay, and finally reaches Axel Heiberg Island. The best part of the trip, over both land and water, will remind you of the rolling, fold-belt uplands seen to the west of Resolute. The sea ice, in its many patterns, is fascinating, and the icecap on south Axel Heiberg has summit levels to 1,400 m (4,600 ft). Along the east side of the icecap lies a tightly folded ridge with sediments dipping at a very steep angle to the west. Within them are the alluvial basins into which the mummified forest remains are interbedded. The approach to Geodetic Hills passes some impressive cliffs that expose the layered sediments; the cliffs also provide nesting sites for raptors such as the gyrfalcon.

Geodetic Hills

On eastern Axel Heiberg Island, not far from Eureka, is the archetypal exposure of a phenomenon now known from several other Arctic sites as a fossil forest, or, more accurately, a mummified

Fossil forest, Geodetic Hills. This 42-million-year-old Metasequoia *(Dawn Sequoia) stump, about 1 meter in diameter at its base, is still carbonaceous and still where it was when a flood buried the forest in sand and gravel. Photograph by B. Howe.*

forest. The remains of these Middle Eocene forests—wood, leaves, seeds, logs, even stumps in situ, their cellulose and lignin still largely intact—have been exposed by fluvial and aeolian erosion in nineteen horizons. Only in these few Arctic sites is it possible to study the details of a 42-million-year-old ecology, and through it the details of the climate of the time. In brief, it appears that thick forests grew in moist valleys or river floodplains at a time when what later became the Princess Margaret range rose rapidly in the west. Periodically, some catastrophic event would cause the forest to become flooded, killing the trees and leaving behind a layer of silt, gravel, and cobbles. The buried forest was anaerobically preserved in water-logged sediments, like Lindall Man, escaping both decomposition and petrification.

The forest at Geodetic Hills was like nothing on earth today. Dominated by no fewer than seven major conifers, among which the Dawn Sequoia (*Metasequoia*) was the most magnificent, the forest flourished in a climate that resembles "Vancouver in the winter, only darker, and Miami in the summer, only brighter." Remarkably the forest grew within a degree or two of its current 80° N latitude, or so specialists in plate tectonics believe. Nowhere on this cooler modern Earth have trees had to adapt to winter darkness or, indeed, summers of constant light.

One geographer, trudging back to the plane, looked back with dismay at his tracks on the hillside and exclaimed that he had done a thousand years' damage in one afternoon. Should sites like these be protected from tourists, scientific or otherwise?

Geodetic Hills to Eureka

DISTANCE: 80 km (50 mi)
FLIGHT TIME: 25 minutes

This trip is a short hop across Eureka Sound. When flying over the northern suburbs of Eureka, watch for dark-brown mudflows with scoop-shaped scars. These are active ground-ice slumps.

Eureka

Eureka on Slidre Fiord is the site of one of five High Arctic weather stations established soon after World War II to increase knowledge of the circulation of Earth's atmosphere and to improve the accuracy of weather forecasting. Observations from the Canadian Arctic were important because much of continental weather is dominated by Arctic air masses. Arctic weather information would also be valuable because transpolar flights were to become common in postwar international travel.

The stations were established jointly by Canada and the United States. Canada provided all the permanent installations and half of the trained personnel, including the officer-in-charge, while the United States provided fuel as well as mobile and other equipment, plus half the staff. The stations at Eureka and Resolute, the first stations to be built in 1947, have collected weather observations continuously since that year. In the early 1970s the United States withdrew its active support of the High Arctic weather stations, and Canada's Atmospheric Environment Service took full control.

The Eureka station reports hourly weather, synoptic weather, upper air soundings, solar radiation, hours of sunshine, precipitation, ice thickness, and freeze-up and break-up. The thirty-year records for Eureka show the mean minimum temperature for February, the coldest month, at −41.4°C (−42.5°F)—the coldest in the Arctic. The mean high temperature in July is 11°C (51.8°F). The coldest recorded temperature was −55.3°C (−67.5°F), and the warmest 19.4°C (66.9°F). On average, 23.4 mm (0.9 in.) of rain and 44.1 cm (17.4 in.) of snow fall each year.

Protected by hills from the north wind, Eureka sits on a low, rolling landscape. Although the area has been glaciated, the surface is made up mainly of weathered bedrock. Because very little vegetation grows, the structures of the underlying geology are evident at the surface. Permafrost produces a series of landforms marked by frost cracks and solifluction. Nearby are some exposures of ground ice—tabular masses of nearly pure ice that have segregated into layers sometimes tens of meters thick. When erosion exposes the ice, it melts back quickly, leaving open amphitheater-shaped scars on the landscape.

Eureka is centrally located in the High Arctic, and the station has become a nodal support point for research scientists. Those with interests in geology and wildlife are especially present during a field season confined to the summer. Recently researchers have examined buried Tertiary forests. They outcrop as layers of accumulated wood and forest litter, alternating with layers of alluvial products. These forests are especially intriguing because they rep-

The Arctic wolf of the Queen Elizabeth Islands has not been subjected to the fanatical persecution that characterizes the interaction between people and wolves in more temperate latitudes and is therefore somewhat more easily seen—and heard. Photograph by DIAND.

resent an unusual opportunity to see a preserved forest floor, with stumps in their original positions and wood, leaves, and seeds in almost mint condition. Paleobotanists, geologists, and foresters are collaborating to unravel the mysteries of how a temperate forest functioned in a polar location with long periods of darkness and continuous summer light.

Another ongoing project involves long-term observation of the hunting and social behavior of Arctic wolves. Now in their sixth field season, researchers have been fortunate to be able to observe animals that have not experienced human predation; consequently, researchers believe, nonverbal "dialogue" can be developed between human and beast. The scientists have taken advantage of some marvelous opportunities to record on film behavior that gives an understanding of the adaptations wolves make to a severe climatic environment and of the ways they organize their space and maintain social order.

⚠ Day Seven

EUREKA, TANQUARY FIORD, LAKE HAZEN, ELLESMERE ISLAND NATIONAL PARK RESERVE, FORT CONGER, ALERT, AND WARD HUNT ISLAND

Eureka to Tanquary Fiord

DISTANCE: 234 km (145 mi)
FLIGHT TIME: 1 hour, 10 minutes

It is an easy run from Eureka over the corner of the Fosheim Peninsula and along Greely Fiord into Tanquary Fiord. Icefields are at a distance on both sides, but active glaciers show their snouts at the water's edge. This area represents textbook geology, with many easily visible folds and faults. On a bright day, we can appreciate why iron oxide is a great coloring agent: the colors of rocks are striking, and whole hillsides are rust, chocolate, or deep raspberry-red. Greely Fiord is, for this latitude, reliably navigable through late summer, although it has icebergs year-round.

Tanquary Fiord

Tanquary Fiord is the northern extension of Greely Fiord, the major reentrant cutting into Ellesmere Island from the west. In 1962 the Defence Research Board began a decade of scientific

Eureka to Alert, to Eureka, to Resolute
Resolute to Iqaluit, to Ottawa

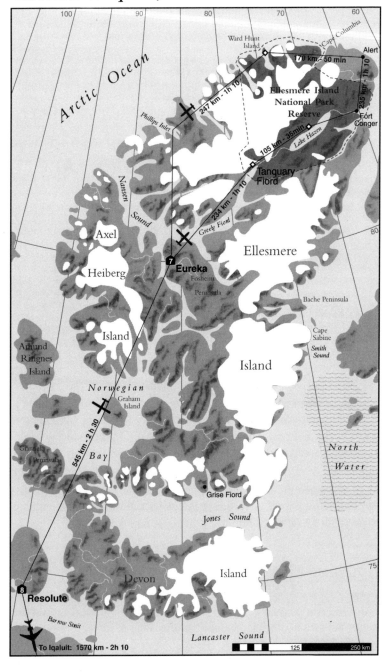

study to establish baseline data about the geology, botany, and zoology of the region. The Canadian icebreaker *Sir John A. MacDonald* delivered a small research team early in August of that year. Subsequently, a small airstrip was cut across the older part of the delta of the MacDonald River, permitting easier access in later years. Tanquary Fiord is about as far north, on the west side of Ellesmere Island, that resupply shipping is possible, and that access is part of the reason that Eureka and Tanquary were chosen as campsites.

The campsite itself is in a beautiful valley. High mountains with valley glaciers that reach the sea, vividly colored rocks, and turquoise water misty with glacial flour present an image that delights the eye. Above the alluvial flat of the delta is a raised beach on which old cultures have left their marks; some Paleo-Eskimo sites date back nearly four thousand years. Evidence of more than a hundred different species of vascular plants and about the same number of mosses and related plants have been documented in the area. Several species of migratory waterfowl also breed here.

Tanquary Fiord to Lake Hazen

DISTANCE: 105 km (65 mi)
FLIGHT TIME: 35 minutes

On this short journey you can pick out the parallel-ridged mountains trending northeast. Watch for glaciers flowing to the valley. When approaching Lake Hazen, you can see the camp shore on the north side protected by a long, low, structural hogback that resembles a morainic island in the lake.

The several narrow and convoluted valleys connecting Tanquary Fiord and Lake Hazen apparently served as the chief migration route for the first pulse of Thule people who later crossed over to Greenland, where they met the incoming Norse. Eureka Sound and Greely Fiord offered possibilities for travel and hunting not available in adjacent territories, and that route led, by chance, to the only practical route to Greenland.

Ellesmere Island National Park Reserve and Lake Hazen

Today, the Tanquary camp is the headquarters of the Ellesmere Island National Park Reserve, an area of 40,000 square kilometers (15,400 sq mi) that extends to the north end of the island. Within the reserve is a large icecap of coalesced glaciers through which the highest peak on the island—and in eastern North America—Barbeau Peak (2,629 m, 8,625 ft) emerges as a nunatak. (Note: Interestingly, the Inuit word *nunatak* entered the geological and English language lexicons in 1877, to signify a hill or mountatin completely surrounded by glacial ice.) Picking up the basic struc-

At the head of Tanquary Fiord is Parks Canada's seasonal headquarters for the Ellesmere Island National Park Reserve. Buildings of insulated fabric over tubular steel frames are anchored against Tanquary's frequent gales. Photograph by H. Swain.

New skill for the borderless world. In Ellesmere Island National Park Reserve, Inuit ranger trainees learn mountain rescue techniques in preparation for adventure-seeking tourists who will travel to the ends of the earth for a frisson *of well-modulated danger. Note the gyrfalcon droppings in the lower right of the picture. Photograph by Parks Canada.*

tural southwest-northeast alignment is Lake Hazen, 80 km (50 mi) long, lying between two parallel mountain ranges. To the west and right behind the lake, is the Garfield range, 1,600 m (5,400 ft) high.

Hazen, a deep lake of 300 m (1,000 ft), is of scientific interest because the area surrounding it remained unglaciated during the last advance and because it is a biological refuge. Research projects have been centered at Lake Hazen since the early 1960s, and the site is being integrated into a growing tourist trade as a depot for walking-camping expeditions between Lake Hazen and Tanquary Fiord.

Parks Canada has a visitors' center and staff accommodations at its headquarters at Tanquary Fiord. The isolation of the site and expense of reaching it severely limits the tourist population, but about a hundred bona fide tourists visit annually. At least that many scientists working on Ellesmere and nearby islands also pass through. Altogether the park warden and two Inuit assistants face a real challenge in managing one of Canada's newest national parks.

Lake Hazen to Fort Conger

DISTANCE: 235 km (146 mi)
FLIGHT TIME: 1 hour, 10 minutes

Leaving Hazen Camp, the flight crosses to the headwaters of the Ruggles River over a broad upland where you may spot musk ox and wolves. The river flows southeast and cuts into the head of a fiord surrounded by scree slopes and steep cliffs whose summit elevation is 860 m (2,800 ft). This steep-sided canal turns north in the Conybeare Fiord, leading to Lady Franklin Bay. At the end of the bay and on its north side stands Fort Conger.

Fort Conger

Fort Conger is forever associated with the tragic and heroic scientific expedition led by Lieutenant A. W. Greely of the United States in the 1880s. The year 1882–1883 was designated the First International Polar Year, reflecting a remarkably prescient attitude toward science and toward international cooperation in scientific research. Worldwide, fourteen expeditions were launched to study the poles. Greely led a group of twenty-four researchers to northern Ellesmere Island to measure various atmospheric and magnetic phenomena and explore the local geology and biota. However, as no human being had yet reached the North Pole, some among the group of twenty-four grew more eager to explore than to record observations. They made an unsuccessful rush for the pole in April of 1882. Back at the research camp, difficulties arose first when the summer resupply failed to get through. A tough winter followed, but they still managed to make another fruitless foray toward the pole in the spring of 1883. Again no ship had arrived by the end of summer, and finally the entire party broke camp and headed by boat for Cape Sabine, 180 km (119 mi) to the south and close to the Greenland shore across Nares Strait. It was here that the seven survivors were rescued in June of 1884. This saga of danger on sea ice, personality conflict, pseudo-military discipline, misery, and starvation—as well as desperate and fumbling attempts at rescue—reveals how heroism and folly are often bound together.

Fort Conger can be viewed only from the air; the site is not open to visitors. You can still see the three buildings of the Greely camp that Robert Peary rebuilt in 1900.

Fort Conger to Alert

North of Fort Conger, the route follows the coast where occasional cliffs rise to 600 m (2,000 ft). As you near Alert, the land gradually lowers to broader and more rounded hills.

Alert

Alert, named after the command ship of the 1875–1876 Nares Expedition, was established in 1950, the last of the joint Canada–U.S. High Arctic weather stations to be built. It was brought in by air from Thule in Greenland after a series of difficulties with icebreaker penetration to the site. Landing first on the sea ice with skis and wheels, a small advance party arrived in early April to build an ice runway, mostly in –35°C (–31°F) weather. The ice strip was needed for the necessary start-up and resupply of equipment and fuel, but construction of the land strip proved difficult because of equipment failure. An airdrop by an RCAF Lancaster on July 31st ended in tragedy when parachutes fouled the elevator of the aircraft and the aircraft crashed in flames. Nine people died:

the crew of seven, a Canadian ice observer, and C. J. Hubbard of the U.S. Weather Bureau, who was the godfather of the High Arctic Weather Stations project. They are buried at the north end of the airstrip next to a monument marking this sad event.

As a weather station, Alert performs similar functions to those carried out at its sister stations, Eureka and Resolute. Recently an additional laboratory was built to support Canada's research on global warming and circumpolar pollution. Greenhouse gases are continuously sampled, and the samples are exchanged worldwide. Traces of the Chernobyl catastrophe were measured at Alert.

As Canada's northernmost occupied site, and the one closest to the former Soviet Union over the pole, Alert had an obvious advantage over other sites for the purpose of monitoring radio communications for Cold War intelligence activities. In 1956 a military team moved in, and Canadian Forces Station Alert, an interservice base, was established in 1958. The infrastructure at the site mushroomed. Today regular military flights in and out are common, since the base has between 200 and 300 personnel. In the interests of military security, access to Alert is strictly controlled.

The July mean high temperature at Alert is 6.4°C (43.5°F), and the low is 0.7°C (33.3°F). The coldest month, February, has a mean annual high temperature of –29.7°C (–21.5°F) and a mean low of –33.6°C (–28.5°F). The coldest temperature recorded for the "top of the world" was –50.0°C (–58°F); the warmest was 20.0°C (68°F). It is dry at Alert, and precipitation is mostly snow—148.1 cm (58.3 in.) on average per year. Rainfall averages 17.5 mm (0.7 in.) and total annual precipitation is 154.4 mm (6.1 in.).

Alert is on a roughly level but raised foreland at the end of the United States mountain range. Hills inland are 450 to 600 m (1,500 to 2,000 ft) high. One can easily detect the southwest-northeast orientation of the gross geologic structure. Boulder fields and fractured rock make up most of the surface.

Alert to Ward Hunt Island

DISTANCE: 170 km (106 mi)
FLIGHT TIME: 50 minutes

Ward Hunt Island is the northernmost stop on the itinerary. To reach it, you pass Cape Columbia, hardly recognizable as a point of land, but in fact the northernmost land in Canada. On the land side of the aircraft, you pass high hills, mountains, the end of the United States Range, and part of the Innuitian fold belt. You can pick out various sediments and structures. The generally lower ground gives way to cliffs and to rough, high knobs about halfway to Ward Hunt. These knobs are really small mountains with elevations up to 1,100 m (3,600 ft). The absence of vegetation reinforces the harshness of this environment. On the seaward side and below, there is nothing but ice. But ice takes different forms. Soon into the flight you notice an ice surface with ridges like frozen waves, with white crests, and with light blue meltwater troughs—all of it in long, parallel lines trending east to west. This is an ice-shelf that can reach 50 m (160 ft) in thickness. It fringes northern Ellesmere Island and surrounds Ward Hunt Island.

Ward Hunt Island—83°05′N

A small island about 4 km (2 mi) long lies at the top of North America. With the farthest north airstrip, the island has become the staging point for many attempts to travel over the pack ice to the true North Pole, or in some cases to transit the pole to a destination in Russia. The depot site on the edge of the ice shelf is littered with the debris of past expeditions. Several temporary structures stand broken open, containing abandoned food and equipment that is decaying and useless. Efforts are being made to remove the refuse and restore the area to a semblance of its natural state.

Seaward of the island's north shore is the ice-shelf zone where over decades fast ice has accreted through snow accumulation and refreezing of meltwater. Its surface is dirty because it is covered with dust blown from the mainland during summer, and the debris is concentrated at the surface by melting of the surface ice. The shelf is a fixed surface of great thickness against which the Arctic ice pack grinds as it moves west along the coast. From time to time, large pieces of the shelf break away and become the ice-islands that drift in the gyre around the polar basin.

Ward Hunt Island to Eureka

DISTANCE: 247 km (153 mi)
FLIGHT TIME: 1 hour, 10 minutes

The reliable route back to Eureka follows the coast of Ellesmere southwest to Phillips Inlet. This part of the route passes along the margin of the polar pack circulating in the Arctic basin. At Phillips Inlet the route cuts over Ellesmere Island on the western side of the icecap covering the British Empire mountains. The summits of the icecap rise as high as 1,800 m (5,900 ft), and are completely white and very high, making visual flight navigation somewhat risky unless the weather is perfect.

Nansen Sound, Greely Fiord, and Phillips Inlet all receive their share of icebergs, reminding us of the difficulties of marine navigation in the High Arctic.

△ *Day Eight*

EUREKA, RESOLUTE, IQALUIT, GREAT PLAINS OF THE KOUKDJUAK, AND OTTAWA

Eureka to Resolute to Iqaluit to Ottawa

From Eureka to Resolute the route retraces, depending on weather, some of the country you have already seen. At Resolute you trade in the Twin Otter aircraft for seats on First Air's 727 to Iqaluit and Ottawa. The long trip over the Northwest Passage (for the last time) and down the length of Baffin Island is flown at over 10,000 m (33,000 ft), so landscape appreciation is more difficult than what you have become used to at lower altitudes. Nevertheless, the flight passes over another of those extraordinary Arctic oases, the Koukdjuak Plain.

Great Plain of the Koukdjuak

This flat, low-lying region of southwest Baffin Island has recently risen from the sea—so recently and so fast, in fact, as to leave one of the world's few populations of freshwater seals behind in what is now Nettilling Lake. Koukdjuak itself, with its permafrost surface patterns and great vegetational zonation with altitude changes of but a few inches, is the breeding ground for millions of wildfowl, especially snow geese and snowy owls. Caribou graze here seasonally. And where there are geese and caribou, predators are

Twin Otter at Beechey Island. A plane intended for back-country work can be identified at a glance by its "tundra tires"—oversized, low-pressure tires, in this case borrowed from a DC-3. Photograph by H. Swain.

not far behind: Arctic foxes, wolves, the occasional polar bear, and, from time to time now, hunters from Cape Dorset and Pangnirtung.

Nettilling Lake has a large shallow bay to the east where marine sediments give way to shield rock. The bay is named after Charles Camsell, the distinguished geologist who later became deputy minister of Northern Affairs. Naming such features after deputy ministers is clearly a good thing.

EPILOGUE: THE PICKUP TRUCK OF THE ARCTIC

Most of any visitor's travel away from the places that have scheduled air service is likely to be on a DHC-6 Twin Otter. These tough twin-engined aircraft, built by DeHavilland in Toronto and powered by PT-6 turbines from Montréal, are worthy successors to a long line of Canadian-designed bush planes that includes the Noordyne Norseman, the Beaver, and the original single radial-engined Otter. DeHavilland built 844 Twin Otters between 1966 and 1989.

Unhappily, the Twin Otter appears to be the last of the line. DeHavilland, facing bankruptcy in 1974, was taken over by the federal government and in 1986 was sold to Boeing. At the time of this writing, the company is again being sold. During the period of government ownership, capital investment was insufficient to modernize the plant or design a new generation of aircraft. Leadership in the industry passed to the Cessna Corporation of Wichita, Kansas, whose Caravan model, still powered by a single Pratt and Whitney Canada PT-6 turbine, is the only successor in sight. Canada, which once built the fastest supersonic interceptor in the world, can no longer build the reliable bush aircraft on which its scattered northern population depends.

PART THREE

Resources

⌀ Fact Sheet for the Northwest Territories

Area 3,426,320 km^2; **Land** 3,293,000 km^2; **Fresh Water** 133,300 km^2; **Population** (1989) 53,326; **Dene** 8,647; **Métis** 3,045; **Inuit** 18,590; **Others** 23,044; **Births** (1988) 1,555; **Deaths** 220; **Marriages** 222.

Economic Accounts (1989)

Gross Domestic Production (Market Prices)	**$2,026,000,000**
Fur production	4,405,000
Fish harvest	2,613,000
Lumber	1,000,000
Metals	950,857,000
Oil and gas	179,071,000
Arts and crafts (estimate)	15,000,000
Private and Public Investment	
Primary industries	582,600,000
Housing	56,000,000
Retail Trade	296,492,000
Restaurant Sales	24,229,000
Social Security—Transfer Payments (1989)	
Old age security	17,159,000
Social Assistance	32,832,000
Family allowance	8,347,000
Unemployment insurance	20,287,000
Total	**78,625,000**
Per capita	**1,474**

N.W.T. Budget 1991-92—$1.1 Billion

Expenditures
Education—17.5%; Health—15.5%; Public works—10.8%; Municipal and Community Affairs—8.6%; Social Services—8.2%; Housing Corporation—7.4%; Transport—5.6%; Renewable Resources—4.0%; Economic Development and Tourism—3.9%; Supplementary—3.9%; Justice—3.6%; Other programs—2.8%; Personnel—2.7%; Government Services—2.0%; Finance—1.9%; Executive—1.6%

Revenues (Federal Government Share)
Grants—73.9%; Transfer Payments—7.5%; Established Program Financing—1.8%

(Own Source Revenues)
Taxation—10.0%; General Revenues—3.2%; Established Program Financing—3.6%

⚠ Hints to the Traveler

There are several matters that you need to attend to prior to your departure from the south which will make your travels less arduous.

AIR SERVICE

There is scheduled jet service from Ottawa, Montréal, and Yellowknife to Iqaluit and Resolute, the main hubs for travel into the Baffin and the High Arctic. Both Canadian Airlines and First Air fly Boeing equipment on these northern runs. First Air also has scheduled HS-748 service up the north coast of Baffin Island and west from Iqaluit into Kitikmeot. If you are really adventurous and coming from Europe, you can connect via First Air's scheduled flights from Iqaluit to Nuuk with KLM out of Greenland.

First Air also has its own charter service, Bradley Air Services Ltd., based out of Resolute. First Air and Bradley enjoy a home base at Carp Airport in downtown metropolitan Carp, Ontario. Within Canada, their toll-free telephone number is 800-267-1247; from the United States, telephone 613-839-3340.

One of the most famous Twin Otter pilots is Mr. Paddy Doyle, of Bradley Air Services. Mr. Doyle, born in Ottawa in 1949, has flown for twenty-three years, twenty-two of them in the Arctic (although he has also flown all the way to the Strait of Magellan). He has always preferred flights over the tundra, sea ice, and glaciers, and has flown everyone—from scientists to nobility—to the far north.

ACCOMMODATIONS

There are co-operative and other inns in the High Arctic, many owned and run by the Inuit. For information and package rates, contact Arctic Co-Operatives Ltd., Hotel Division, 1741 Wellington Avenue, Winnipeg, Manitoba R3H 0G1, Canada (telephone 204-786-4481). Additional information is also available in the Northwest Territorial Airways inflight magazine, *The Northwest Explorer*. It can be obtained from the Northwest Territorial Airways, P.O. Box Service 9000, Yellowknife, N.W.T. X1A 2R3, Canada (telephone [403] 920-4576).

CLOTHING AND FITNESS

First, it is important that all travelers to the High Arctic be physically fit and in good health.

Your clothing should have several light layers, including a windproof and waterproof outer shell, sufficient to keep you warm at a windchill temperature of −30°C, but that also allows you to shed layers when temperatures are at a more normal summer level of 0°C to 10°C. You should have warm (lined) gloves or mittens, and a knitted cap or hat that covers your ears. Long underwear is also useful. You will need stout boots with Vibram or similar soles, and uppers that will keep out water, mud, and snow. Thick, warm socks are needed as well.

On the trip you will also need a light, preferably down, sleeping bag and your own towel and washcloth.

Do not forget sunglasses, preferably with polarized lenses. The glare from snow-lowered surfaces can literally be blinding, and you will have no respite at night.

MEDICINES

Bring aspirin and (just in case) an anti-diarrhetic. If you have to use prescription drugs, bring all that you will require plus the prescription. Recreational drugs are not legal.

METHODS OF PAYMENT

Although credit cards (especially Visa) are accepted at most places in the Arctic, there are exceptions. It is best to have Canadian currency or travelers cheques available as well.

Canada has a 7 percent value-added tax known as the GST. Non-residents can be reimbursed for this tax by filling out a brief form, available from Customs, and providing receipts.

PHOTOGRAPHIC EQUIPMENT

Your camera should not be too fragile. A fast zoom telephoto lens of medium focal length will probably become your lens of choice; things are always farther away than they seem in the clear Arctic air. Be careful about using too fast a film, however, as continuous sunshine and glare from ice and snow can otherwise leave you shooting at tiny apertures and extreme speeds.

TOURS

Package tours are available. Contact Baffin Tourism Association, Box 820, Iqaluit, Northwest Territories X0A 0H0, Canada.

TRAVEL DOCUMENTS

If you are coming from the United States, you will need a birth certificate or a passport. From many other countries, in addition to your passport, you will need a visa. Check with the Canadian embassy or consulate in your country before departure.

WEATHER AND WHEN TO TRAVEL

The preferable time of travel is during mid- to late summer. But travel schedules can be affected at any time, the result of fog, white-outs, or storms.

◬ Suggested Readings

The books listed below are books that we admire and that are up-to-date. They will provide the traveler with a good overview of the High Arctic and, it is hoped, many hours of pleasurable and informative reading.

Berton, Pierre. *The Arctic Grail: The Quest for the Northwest Passage and the North Pole, 1818–1909.* Toronto: McClelland and Stewart, 1988.
  The master storyteller takes us from the newly unemployed British navy of 1815 to the final triumph of Peary and Henson in 1909. For the Royal Navy hardies of the nineteenth century, desperate adventures provided object lessons in the costs of ignoring local climates and customs in the search for a passage of no discernible commercial or military value. Also in paperback.

Brody, Hugh. *Living Arctic: Hunters of the Canadian North,* Toronto: Douglas and McIntyre, 1987.
  Brody looks at his subject from the inside. An excellent observer and sensitive writer, he gives us the realities of being Inuit not in the past, but today.

Canada. *Canada's North: The Reference Manual.* Rev. ed. Ottawa: Department of Indian Affairs and Northern Development, 1990.
  This periodic compendium of facts, figures, and maps covers most topics in fourteen terse chapters, each with a bibliography. It is available in the better Canadian bookstores or by mail from Supply and Services Canada, Ottawa K1A 0S9.

Damas, D., ed. *Handbook of North American Indians.* Vol. 5, *Arctic.* Washington D.C.: Smithsonian Institution, 1984.

Not much is left out of this volume. It covers all of the North American Arctic aboriginal world and is the authority on prehistory to the present. A wonderful reference, but not exactly bedtime reading.

Grant, Shelagh D. *Sovereignty and Security? Government Policy in the Canadian North 1936–1950.* Vancouver: University of British Columbia Press, 1988.

This scholarly volume provides about the only comprehensive background for understanding the evolution of official attitudes toward the Arctic.

Lopez, Barry. *Arctic Dreams: Imagination and Desire in a Northern Landscape.* New York: Scribner's 1986.

Written by the ultimate romantic. A wonderful, breathtaking look by a traveler of imagination and ironic perception. Also available in paperback.

McGhee, Robert. *Canadian Arctic Prehistory.* Ottawa: Canadian Museum of Civilization, 1990.

This recent, reliable, and very readable account of Inuit prehistory is by one of the most respected authorities in the field.

Northwest Territories Data Book 1990/91. Yellowknife: Outcrop, Ltd., 1990.

Published annually, the *Data Book* is the most complete collection of facts about the Northwest Territories, covering the physical character of the land, the resources, the society, and the economy. It is an "all you need to know" book, and worth using to keep up to date.

Phillips, David. *The Climates of Canada.* Ottawa: Supply and Services Canada, 1990.

In an excellent and easy guide through the climates and elements, the author explains Canadian climate by province. Two separate chapters on Yukon and Northwest Territories include unusual and valuable detail about the cold, dry Arctic land and frozen sea.

Williams, P. J. *Pipelines and Permafrost: Science in a Cold Climate.* Ottawa: Carleton University Press, 1986.

A slender volume, but nevertheless packed with general information about permafrost and how it affects natural landscapes and human works. We have much to learn about the frozen earth as experiments with pipelines reveal.

Index

Alert, 135–136
alluvial basins, 123
amauti, 58
Anglican church, 36
archaeology, 22, 97
Arctic, 4
Arctic Bay, 97
Arctic Platform, 7
art, 51
Assistance Bay, 113
Axel Heiberg Island, 123

Baffin, William, 28, 77
Baffin Bay, 19
Baffin Island, 7
Baker Lake, 52
Banks Island, 8
Beechey Island, 113
Bent Horn, 110–111
birds, 109
Blacklead Island, 78
boats, 57
Boothia Peninsula, 7
boreal forest, 4
Broughton Island, 82
buildings, 58
Bylot Island, 88

Camsell, Charles, 140
Canadian Shield, 7
Canadians, 60
Cape Columbia, 137
Cape Dorset, 52
caribou, 23

carving, 51, 53
Catholic church, 36
Cessna Corporation, 141
Christianity, 35
churches, 36
Clyde River, 52, 87
colonization, 119
Cornwallis Island, 103
crafts, 51
Craig Harbour, 118
Cumberland, 78
currents, 17

Davis, John, 77
death, 41
DeHavilland Corporation, 141
Department of Indian Affairs and
  Northern Development, 64
des Groseilliers, Médard Chouart,
  28
Devon Island, 116
DEW Line, 37
disease, 39, 48
diversity, 75
dolomite, 8, 108
Dorset, culture, 24
Dorset, people, 24
drainage, 108
Dundras Harbour, 118

Eclipse Sound, 93
education, 42
elevations, 7, 8
Ellesmere Island, 7

Ellesmere Island National Park Reserve, 132
employment, 49
environment, 61
equinox, 9
Erebus, 29, 115
erosion, 124
Eureka, 125
Eureka Fiord, 19
exploration, 28–29, 31

fertility, 41
First International Polar Year, 134
fog, 13
forests, middle Eocene, 124
Fort Conger, 134
Foxe, Luke, 28
foxes, arctic, 140
Franklin, John, 115
Frobisher, Martin, 28, 71
Frobisher Bay, 72
fulmars, 99

Geodetic Hills, 123
geology, 5, 105, 107, 129
glaciation, 77
glaciers, 84
gneiss, 7
government, 36, 54
granite, 7
Greely, A. W., 134
Grise Fiord, 117
guillemots, 99
gulls, 99
gyres, 18

hares, Arctic, 109
health, 46, 83
Hearne, Samuel, 29
Hetherington, Charles, 111
history, 117
houses, 56
housing, 44

Houston, James, 51
Hubbard, C. J., 136
Hudson, Henry, 28
Hudson Bay, 7
Hudson's Bay Company, 29

iglu, 23
Ikpiarjuk, 97
Independence I, 22
insurance, medical, 48
Inuit, 3, 21, 39
Inukjuak, 121
Inuktitut, 71
Inuvialuit Final Agreement, 65
inversions, 13
Iqaluit, 71, 73–74
Isabella Bay, 85

James, Thomas, 28
jobs, 49

Keewatin District, 10
Kekerten Island, 78, 79
King George V Mountain, 98
kittiwakes, 99
komatik, 58
kooneak, 98

Labrador Current, 18
Lake Harbour, 52
Lake Hazen, 133
Lancaster Sound, 19
land, 64
landforms, 8
landscapes, 69
lead, 106
lead poisoning, 115
limestone, 8, 99, 108
Lindstrom Peninsula, 119
litter, 115

machines, 57
Mackenzie, Alexander, 29

Index

Mackenzie Basin, 11
Mackenzie Valley, 10
Magnetic North Pole, 111
mammals, 110
mammals, marine, 107
Manhattan, 63
M'Clintock, Leopold Francis, 31, 115
meadowlands, 116
Melville Peninsula, 7
migration, 131
mines, 50, 95
mineralization, 106
missions, 35
moraines, 73
Munk, Jens, 28
murres, 99

Nanisivik, 94, 97
Nanisivik Mines Ltd., 95
Nares Expedition, 135
narwhals, 93
Nettilling Lake, 139
North Water, 19
Nunatak, 132
Nunavut, 66

oil, 112
ore, 95, 106
out-migration, 121

Pangnirtung, 52, 77
Pangnirtung Fiord, 77
Parks Canada, 134
Parry, W. E., 101
Parry Channel, 19
Peary, Robert E., 62
Peck, E. J., 35
Penny, William, 77
permafrost, 14–15
plate tectonics, 125
Polar Bear Pass, 107
Polaris Mine, 106
Polar Sea, 63

polynyas, 19
Pond, John, 89
Pond Inlet, 89
population, 41
Povungnituk, 52
Prince Leopold Island, 99
pyrite, 95

Radisson, Pierre, 28
Rae, John, 31
rainfall, 12
relief, 73
relocation, 119
rescue, 135
research, 83
Resolute, 101
Resolute Bay, 102
resources, natural, 50
roads, 56
Roman Catholic church, 36
Ross, James Clark, 111
Ross, John, 77, 89

schools, 38, 42
seabirds, 90
sea ice, 14, 21
sealing, 82
seals, 95
services, social, 76
settlement, 38, 61
shales, 8
shorebirds, 94
snow, 12
snowmobiles, 57
soapstone, 52
solstice, 9
storms, cyclonic, 11
summer, 11, 13
Sverdrup, Otto, 31, 118

Tanquary Fiord, 129
temperature, 10, 11, 74
Terror, 29, 115

Thule, 27
Thule, people, 25
Titanic, 18
tool, 22, 25
trading, fur, 34
tree line, 5
treaties, 65
tuberculosis, 47
tundra zone, 5
Tungavik Federation, 65
Twin Otter, 141

umiak, 25
urbanization, 38
"utilidors," 45

vegetation, 4, 102, 108

Victoria Island, 8
Vikings, 25

walruses, 107
Ward Hunt Island, 137
weather forecasting, 125
whales, 31, 85, 93
whaling, 31, 85
whiteout, 13
wildfowl, 139
wind, 10
winter, 9, 12
wolves, Arctic, 127
woodland, 4
work, 49

zinc, 106